Lecture Notes in Artificial Intelligence 11976

Subseries of Lecture Notes in Computer Science

Series Editors

Randy Goebel
University of Alberta, Edmonton, Canada
Yuzuru Tanaka
Hokkaido University, Sapporo, Japan
Wolfgang Wahlster
DFKI and Saarland University, Saarbrücken, Germany

Founding Editor

Jörg Siekmann
DFKI and Saarland University, Saarbrücken, Germany

More information about this series at http://www.springer.com/series/1244

Peipeng Liang · Vinod Goel ·
Chunlei Shan (Eds.)

Brain Informatics

12th International Conference, BI 2019
Haikou, China, December 13–15, 2019
Proceedings

 Springer

Editors
Peipeng Liang
Capital Normal University
Beijing, China

Vinod Goel
York University
Toronto, ON, Canada

Chunlei Shan
Shanghai University of Traditional
Chinese Medicine
Shanghai, China

ISSN 0302-9743   ISSN 1611-3349   (electronic)
Lecture Notes in Artificial Intelligence
ISBN 978-3-030-37077-0   ISBN 978-3-030-37078-7   (eBook)
https://doi.org/10.1007/978-3-030-37078-7

LNCS Sublibrary: SL7 – Artificial Intelligence

This Springer imprint is published by the registered company Springer Nature Switzerland AG
The registered company address is: Gewerbestrasse 11, 6330 Cham, Switzerland

Preface

The International Conference on Brain Informatics (BI) series has established itself as the world's premier research conference on Brain Informatics, which is an emerging interdisciplinary and multidisciplinary research field with the vision of investigating the brain from the informatics perspective. Firstly, Brain Informatics combines the efforts of Neuroscience, Cognitive Science, Machine Learning, Big Data Analytics, Artificial Intelligence (AI), and Information and Communication Technology (ICT) to study the brain as a general information processing system. Secondly, new informatics equipments, techniques, and platforms are causing a revolution in understanding the brain. Thirdly, starting from its proposal as a field, the goal of Brain Informatics is to inspire future AI, especially Web Intelligence (WI, i.e. AI in the connected world). BI 2019 provided a premier international forum to bring together researchers and practitioners from diverse fields for presentation of original research results, as well as exchange and dissemination of innovative and practical development experiences on Brain Informatics. The main theme of BI 2019 is "Brain Science meets Artificial Intelligence" with respect to the five tracks: Cognitive and Computational Foundations of Brain Science; Human Information Processing Systems; Brain Big Data Analytics, Curation and Management; Informatics Paradigms for Brain and Mental Health Research; and Brain-Machine Intelligence and Brain-Inspired Computing.

The BI conference series started with the WICI International Workshop on Web Intelligence meets Brain Informatics, held in Beijing, China in 2006. The 2nd, 3rd, 4th, and 5th BI conferences were held in Beijing, China (2009); Toronto, Canada (2010); Lanzhou, China (2011); and Macau, China (2012), respectively. Since 2013, health was added to the conference title as Brain Informatics and Health (BIH) with an emphasis on real-world applications of brain research in human health and well-being. The BIH 2013, BIH 2014, BIH 2015, and BIH 2016 were held at Maebashi, Japan; Warsaw, Poland; London, UK; and Omaha, USA, respectively. In 2017, the conference went back to its original design and vision to investigate the brain from an informatics perspective and to promote a brain-inspired information technology revolution. Thus, the conference name was changed back to Brain Informatics in Beijing, China, in 2017. In 2018, the conference was held in Arlington, Texas, USA. In 2019, this grand event was held in Haikou, China, during December 13–15. The BI 2019 conference was hosted by Hainan University, and officially sponsored by the Chinese Association for Artificial Intelligence (CAAI) and Web Intelligence Consortium (WIC).

BI 2019 solicited high-quality papers and keynote talks with world-class speakers, a panel discussion, workshops, and special sessions. BI 2019 involved several world leaders in brain research and informatic technologies, including Aaron Ciechanover, Lin Chen, Qionghai Dai, Hongkui Zeng, Michael Fox, Hesheng Liu, Jing Luo, Tianzi Jiang, Yong He, and many other outstanding researchers. These proceedings contain 26 high-quality papers accepted and presented at BI 2019, which bridges scales that span from atoms, to thoughts and behavior, and provides a good sample of state-of-the-art research advances on Brain Informatics from methodologies, frameworks, techniques, to applications and case studies.

The BI 2019 conference promoted more multidisciplinary studies on innovative and integrated solutions to improve our ability to understand the brain, and promoted more brain-inspired technologies and BCI systems for important real-world applications. In recent years, new experimental methods, such as neuroimaging, opto-genetics, and dense-electrode recording are generating massive amounts of brain data at very fine spatial and temporal resolutions. The new data fusion and AI methodologies are welcomed to enhance human interpretive powers when dealing with big neuroimaging data, including fMRI, PET, MEG, EEG, and fNIRS, as well as from other sources like eye-tracking and from wearable, portable, micro, and nano devices. Brain Informatics research creates and implements various tools to analyze all the data and establish a more comprehensive understanding of human thought, memory, learning, decision-making, emotion, consciousness, and social behaviors. These methods and related studies will also assist in building brain-inspired intelligence, brain-inspired computing, human-level wisdom-computing paradigms and technologies, and improving the treatment efficacy of mental health and brain disorders.

We would like to express our gratitude to all BI 2019 Conference Committee members for their instrumental and unwavering support. BI 2019 had a very exciting program with a number of features, ranging from keynote talks to a panel discussion, workshops, and special sessions. This would not have been possible without the generous dedication of the Program Committee members in reviewing the submitted papers and abstracts, the BI 2019 workshop and special session chairs and organizers, and our keynote speakers in giving outstanding talks at the conference. BI 2019 could not have taken place without the great team effort of the Local Organizing Committee and generous support from sponsors. Particularly, we would like to show our sincere apperication to our kind sponsors, including Hainan University, Chinese Association for Artificial Intelligence, Web Intelligence Consortium, Hainan Provincial Health Committee, Department of Commerce of Hainan Province, Haikou Municipal Health Committee, Chinese Society for Cognitive Science, Chinese Psychological Society, CAAI Technical Committee on Brain Science and Artificial Intelligence, Chinese Research Hospital Association (CRHA) Medical Imaging and Artificial Intelligence Branch, Chinese Neuromodulation Society, Chinese Minister of Science and Technology (Dalian Talent Bank, Foreign Talent Resources), IEEE Computational Intelligence Society, International Neural Network Society, Springers *Lecture Notes in Artificial Intelligence*, Hainan Medical University, Hainan Province Enterprise Directors Association, Hainan Province Enterprise Confederation, Haikou Convention & Exhibition Industry, Beijing Qingyang Science and Technology Co., Ltd., Beijing 7invensun Technology Co., Ltd., Beijing LongCheng Health Big Data Technology Co., Ltd., Beijing LongCheng Virtual Business Investment & Management Co., Ltd., Beijing LongCheng Natural Science Research Institute, Institute of People's Health, Beijing LongCheng Quantum Technology Research Institute, Chang Zhi ASCN Medicine Group Co., Ltd., Yellow River Bank Preparatory Committee, Zhong Nan Bank Preparatory Committee, Zhong Yu Wealth (Beijing) Cultural Communication Co., Ltd., Selfwealth (Beijing) Technology Co., Ltd., Wuhan OE-Bio Co., Ltd., and Neuracle Technology (Changzhou) Co., Ltd.

Special thanks go to the Steering Committee co-chairs, Qionghai Dai, Ning Zhong, and Hanchuan Peng, for their help in organizing and promoting BI 2019. We also thank Juzhen Dong for his assistance with the CyberChair system. We are grateful to Springer *Lecture Notes in Computer Science* (LNCS) for their sponsorship and support. We thank Springer for their help in coordinating the publication of this special volume in an emerging and interdisciplinary research field.

December 2019

Peipeng Liang
Vinod Goel
Chunlei Shan

Organization

General Chair

Qingming Luo   Hainan University, China

Program Committee Chairs

Peipeng Liang   Capital Normal University, China
Vinod Goel   York University, Canada
Chunlei Shan   Shanghai University of Traditional Chinese Medicine, China

Organizing Chairs

Xin Lou   Chinese PLA General Hospital, China
Nan Ma   Beijing Union University, China
Feng Xu   Tsinghua University, China

Workshop/Special Session Chairs

Yang Yang   Beijing Forestry University, China, and MAEIT, Japan
Mufti Mahmud   Nottingham Trent University, UK

Publicity Chairs

Zhiqi Mao   Chinese PLA General Hospital, China
M. Shamim Kaiser   Jahangirnagar University, Bangladesh

Steering Committee Co-chairs

Qionghai Dai   Tsinghua University, China
Ning Zhong   Maebashi Institute of Technology, Japan
Hanchuan Peng   Allen Institute for Brain Science, USA

Program Committee

Jun Bai   Institute of Automation, Chinese Academy of Sciences, China
Weidong Cai   The University of Sydney, Australia
Lihong Cao   Communication University of China, China
Mirko Cesarini   University Milano-Bicocca, Italy
Jianhui Chen   Beijing University of Technology, China

Denggui Fan	University of Science and Technology Beijing, China
Yong Fan	University of Pennsylvania, USA
Liheng Bian	Beijing Institute of Technology, China
Przemyslaw Biecek	University of Warsaw, Poland
Nizar Bouguila	Concordia University, Canada
Changquan Long	Southwest University, China
Feiyan Chen	Zhejiang University, China
Xun Chen	University of Science and Technology of China, China
Gopikrishna Deshpande	Auburn University, USA
Hongwei Dong	University of Southern California, USA
Lu Fang	Tsinghua-Berkeley Shenzhen Institute, China
Mohand-Said Hacid	Universite Claude Bernard Lyon 1, France
Bin He	Carnegie Mellon University, USA
Hongjian He	Zhejiang University, China
Tianzi Jiang	Institute of Automation, Chinese Academy of Sciences, China
Colin Johnson	University of Kent, UK
Yongjie Li	University of Electronic Science and Technology of China, China
Youjun Li	North China University of Technology, China
Feng Liu	Harvard Medical School, USA
Ke Liu	Chongqing University of Posts and Telecommunications, China
Tianming Liu	University of Georgia, USA
Weifeng Liu	China University of Petroleum, China
Yiguang Liu	Sichuan University, China
Lucelene Lopes	Pontifical Catholic University of Rio Grande do Sul, Brazil
Roussanka Loukanova	Stockholm University, Sweden
Gang Pan	Zhejiang University, China
Abdel-Badeeh Salem	Ain Shams University, Egypt
Divya Sardana	University of Cincinnati, USA
Dominik Slezak	University of Warsaw, Poland
Neil Smalheiser	University of Illinois, USA
Diego Sona	Istituto Italiano di Tecnologia, Italy
Niels Taatgen	University of Groningen, The Netherlands
Ryszard Tadeusiewicz	AGH University of Science and Technology, Poland
Xiaohui Tao	University of Southern Queensland, Australia
Egon L. Van den Broek	Utrecht University, The Netherlands
Xiaohong Wan	Beijing Normal University, China
Changdong Wang	Sun Yat-sen University, China
Guoyin Wang	Chongqing University of Posts and Telecommunications, China
Junkai Wang	Tsinghua University, China
Qing Wang	Northwestern Polytechnical University, China

Contents

Brain-Machine Intelligence and Brain-Inspired Computing

Special Session on Computational Social Analysis for Mental Health

Cognitive and Computational
Foundations of Brain Science

EEG Signal Indicator for Emotional Reactivity

Guodong Liang[1], Xiangmin Xu[1(✉)], Zicong Zheng[1], Xiaojie Xing[2],
and Jianxiong Guo[3(✉)]

[1] South China University of Technology, Guangzhou 510006, China
xmxu@scut.edu.cn
[2] Guangdong Brainview Intelligent Technology Co., Ltd.,
Foshan 528200, China
[3] Guangzhou Huiai Hospital, Guangzhou 510370, China
jxguonet01@126.com

Abstract. The main purpose of this study was to seek indicators that could effectively describe an individual's subjective emotional reactivity with objective physiological signals. The study used VR to present two valences of stimuli to induce certain emotions, meanwhile collecting multimode physiological data. Then three dimensions of emotional reactivity, namely emotional intensity, emotional sensitivity and emotional persistence, measured by a self-assessment questionnaire, were correlated with individual's EEG signals. Results showed that emotional persistence was significantly positively correlated with the change rate of both alpha bands under HVHA stimulus and beta bands under LVHA stimulus between two cerebral hemispheres. These findings are in accord with the frontal EEG lateralization theory that the left and right prefrontal cortex hemispheres process differently for multiple emotions. The change rate in alpha/beta ratio of the left hemisphere between the baseline and after HVHA stimulus was also found significantly positively correlated with emotional reactivity along with the sensitivity and intensity dimensions. This indicate that the relative power spectrum ratio of two frequency bands is more effective.

Keywords: Emotional reactivity · EEG signals · VR scene

1 Introduction

With the continuous development of neuroscience, researchers have found that there is a physiological basis for emotional response. Gross et al. [1] indicated that there are two parallel physiological systems closely related to personality in human brain, namely behavioral inhibition system and behavioral activation system. These two systems control individual response to all kinds of stimulus and also manifest itself during emotion reactivity. Later, researchers [2] found that the prefrontal cortex is the neural basis for the emotional response. Accordingly, more and more researchers believed that there is a stable difference of emotional response among individuals. For example, Rothbart's results suggested that emotional response is an important dimension of temperament [3]. They believed that, in essence, temperament reflect the difference about individual reaction and self-regulation in terms of emotion and attention.

© Springer Nature Switzerland AG 2019
P. Liang et al. (Eds.): BI 2019, LNAI 11976, pp. 3–12, 2019.
https://doi.org/10.1007/978-3-030-37078-7_1

Different views have been proposed by researchers on the characteristics of emotional response. Larsen et al. [4] argued that the intensity of emotional experience is an important aspect of emotional response. Davidson [5, 6] thought that the differences in emotional responses among individuals can be manifested by emotional activation thresholds, peaks, time to peak, and recovery time. Linehan [7] described the emotional response of patients with borderline personality disorder to negative emotional stimuli as a sensitive group with high emotional intensity and a slow return to emotional baseline levels.

Summarizing the predecessors' work, Nock et al. [8] proposed the concept of "emotional reactivity" to describe the differences of the characteristics in emotional responses among individuals. They refer emotional reactivity (ER) to individual (1) emotional experience (ie, emotional sensitivity) to a range of stimuli, and (2) the intensity of the experienced emotion (ie, emotional intensity), and (3) the length of time (ie, emotional persistence) required for the mood to return to the baseline level.

The proposition of this concept distinguished emotional responsiveness from previous related concepts such as specific emotions (eg, anger, sadness, and anxiety), temperament, and emotional variability. It provided a new stimulus-response process that is easy to measure [9]. Therefore, more and more researchers are now trying to understand and assess the differences of emotional responses among individuals using this concept.

In this work, in order to estimate human emotion reactivity and establish an association between emotional reactivity and individual physiological signals, virtual reality (VR) scenes were used as emotional stimulus to trigger one's certain emotions and multiple physiological signals were collected.

The conventional methods basically utilized visual and audio attributes to awaken human emotion, such as picture, music, and movie clips. But with the recent technological advances of VR, a new way to induce emotion has come up to people's field of vision. VR is a virtual reality technology, which provides comprehensive use of computer technology and graphics technology. At the same time, through different control devices and display devices, it can form an immersive feeling in the computer. Due to the high quality of immersion and visualization, VR stimulus are different and to some degree superior for emotion induction to the traditional ones [10]. Nowadays, researchers have broaden its application in many areas, such as driving simulators [11], architecture design [12] and medical training [13] et al.

Affective computing refers to the acquisition of facial or physiological changes caused by human emotions through various sensors, and the use of emotional models to identify and analyze these signals to understand human emotions and respond appropriately [14]. At present, the researches on affective computing mainly focuses on the distinguishability of affect through analysis of EEG, EMG and autonomic nervous systematic activities and so on [15–17]. But exploring under what conditions and which emotions could obtain distinguished physiological activities are more meaningful than seeking a constant relationship between emotional experience and physiological response.

The objective of this study presented in the paper is to systematically uncover the association between different dimension of emotional reactivity and multimode physiological signals under certain type of emotion.

2 Methodology

2.1 Experiment Setup

The experiment contained two sessions, a preliminary experiment and a formal experiment. It was conducted in a laboratory environment with controlled illumination and temperature. Once arrived, the participants received a short oral briefing about the experiment and signed an informed consent before the experiment proper starts. Then, they were asked to finish a series of questionnaire, which included GAD-7, PHQ-9 and Emotional Reactivity Scale (ERS). After that, participants were attached to neuro-feedback and biofeedback system (NeXus-10, Germany) and also a wristband named Empatica E4 to record their physiological data. Figure 1 shows the experiment design. Data from sensors are collected during the procedure.

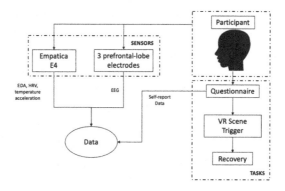

Fig. 1. Experiment design

2.2 Procedure

Two VR scenes out of six selected from former study [18] were used in the formal experiment. They each represented a HVHA stimulus and a LVHA stimulus [19]. One scene was selected in pre-experiment [20]. All stimuli were shown in Table 1. The scenes were played through HTC VIVE.

Table 1. Name, valence and arousal of each VR scene

No.	Name	Valence	Arousal
1	Farm with animals	7.5	6.2
3	Dark basement	1.9	8.4
2	Villa tour	6	3.9

The whole procedure lasted about 15 min. For each VR scene, it contained four phases, a resting phase, a stimuli phase, a recovery phase and a diversion phase. During the resting phase, which lasted 30 s, participants were asked to try to rest themselves and stay still. Then with a sound beep as a cue to remind the participants the second phase began. In this phase, participants were asked to explore the VR environment all the way they wanted with a hint that there would be a time bar appearing in the last 10 s. This was to remind the participant that the VR scene would be ended in 10 s, please be ready to get into the next phase and chose a comfortable position to remain still. The third phase last 60 s for the participants to calm themselves down from the former VR scene. Researches proved that it takes about 2 to 4 min for human to recover from an emotional arousal state, so a diversion task which is the most efficient emotion regulation strategy was chosen for the last phase that contained two calculation tasks [21]. Each task lasted 10 s for the participant to respond. Figure 2 shows the whole process of each scene. The pre-experiment was to help participants get familiar with the process.

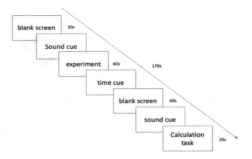

Fig. 2. Flow diagram of the experiment

2.3   Participant

There were 22 participants took part in this study, all of whom were students from South China University of Technology. They aged from 22 to 25 years old and were composed of 12 males and 10 females. All participants were fluent in Chinese with no deformities and were asked to have a good rest on test eve. Each subject signed a consent form and was given a brief instruction about the meaning and protocol of the experiment before beginning.

2.4   Measures

During the whole procedure, physiological responses were measured with neurofeedback and biofeedback system (NeXus-10, Germany) and Empatica E4. EEG signals was recorded with a 3 prefrontal-lobe electrodes lead [22], EDA, HRV, temperature and acceleration were recorded by E4.

Mental health state was evaluated according to the GAD-7 and PHQ-9.

Emotional reactivity was assessed by Emotional Reactivity Scale (ERS) [8]. ERS is an important tool for measuring individual emotional reactivity, which includes emotional sensitivity, emotional intensity, and emotional persistence three subscales. ERS has been verified among clinical and college researches [8, 23, 24], and has been widely used in the world. Researchers from the Netherlands, France, China and other places have localized and revised this questionnaire [23, 25, 26].

3 Result

The self-assessment of mental health state was to rule out inappropriate subjects. GAD-7 with 7 items was used to assess subject's stress and 1 subject was removed because of his score of GAD-7 is 11, the rest were all included within the next analysis (1.75 ± 1.36). PHQ-9 with 9 items was used to assess their depression symptom, and no subject was out (2 ± 1.04).

Emotional reactivity was measured by ERS, whose result was showed in Table 2. Correlation analysis indicated the three dimensions of emotional reactivity are significant related. An independent Sample T test that was conducted to verify whether gender affects emotional reactivity revealed no significant difference between male and female in respect of emotional sensitivity, emotional intensity, emotional persistence and emotional reactivity.

Table 2. Score of male and female in ERS

	Gender	M ± SD
Emotional sensitivity	Male	20.22 ± 10.26
	Female	20.20 ± 8.44
Emotional intensity	Male	14.00 ± 5.83
	Female	13.80 ± 6.38
Emotional persistence	Male	8.44 ± 3.24
	Female	8.00 ± 3.94
Emotional reactivity	Male	42.67 ± 18.67
	female	49.67 ± 21.36

For the investigation of the association between emotional reactivity and EEG signals, the EEG data were commonly average referenced, downsampled to 250 Hz, and high-pass filtered with a 2 Hz cutoff frequency using Python (SCIPY). The EEG signal embedded with various artifacts such as eye artifacts, muscle artifacts, and power line source. In this study, EEG signal was recorded with 3 prefrontal-lobe electrodes lead mainly focused on EEG frequency alpha (8–12 Hz) and beta (12–30 Hz) bands, hence the influence of eye artifacts which is dominant below 4 Hz, muscle artifacts above 30 Hz and power line artifacts above 50 Hz are eliminated [27].

Former researches concluded that the interaction between human and the VR environment could be an important factor to improve immersion and effectiveness [28], so participants were asked to explore the VR environment as possible as they could to make the most of the stimulus. The 5 s of EEG signal before each stimulus phase, namely the resting phase, were extracted as baseline. The first 10 s and last 5 s of EEG signal after each VR scene were extracted for further analysis for the dynamics of affective responding [21].

The change rate of the ratio of alpha and beta bands between two cerebral hemi-spheres under different conditions in the recovery phase was calculated for the reason of the functional differentiation of human brain which means that the activation responses under different type of emotional stimuli of the right and left cerebral hemispheres are diverse. Bivariate correlational analyses were used to determine the relationship between emotional persistence and these change rates. A pearson corre-lations was conducted using a 2000 sample bootstrap. As shown in Table 3, emotional persistence is significantly positively correlated with change rate of alpha bands under HVHA stimulus, which is also significantly positively correlated with emotional reactivity, and change rate of beta bands under LVHA stimulus. This may suggest that the asymmetry activity of the alpha bands in the frontal lobe is related to the individual ability to maintain and regulate emotions, which is consistent with the findings of Moïra Mikolajczak [29].

Table 3. Correlations between emotional reactivity scores and change rate of alpha and beta bands

	HVHA				LVHA			
	Change rate of alpha bands		Change rate of beta bands		Change rate of alpha bands		Change rate of beta bands	
	r or [95% CI]	p	r or [95% CI]	p	r or [95% CI]	p	r or [95% CI]	p
Emotional persistence	.636[.024, .907]	.026*	−.281[−.725, .396]	.376	−.095[−.646, .558]	.769	.584[.158, .922]	.046*
Emotional reactivity	.662[.137, .938]	.019*	−.141[−.680, .544]	.662	−.368[−.784, .336]	.240	.480[−.444, .860]	.114

N = 21, all correlations assessed two-tailed using a 2000 sample bootstrap with bias corrected accelerated confidence intervals.

Since the signal-to-noise ratio of EEG signals was relatively low, the specificity of the parameters is not strong enough when single frequency band was used independently, neither was the sensitivity [30]. Therefore, many scholars began to use the relative power spectrum ratio of two frequency bands. The change rate of the alpha/beta bands between the baseline and the first 10 s of the recovery phase were calculated. Another bivariate correlational analysis was conducted among the change rate, emotional sensitivity, emotional intensity, and emotional reactivity. Table 4 shows that the rate of change in alpha/beta ratio of the left hemisphere between the baseline and after HVHA stimulus is significantly positively correlated with emotional reactivity, sensitivity, and intensity. This may open the possibility that the faster a person's mental state changes, the stronger the emotional response, and the more emotional feelings can be perceived.

Table 4. Correlations between emotional reactivity scores and change rate of alpha/beta bands

	Emotional reactivity		Emotional sensitivity		Emotional intensity	
	r or [95% CI]	p	r or [95% CI]	p	r or [95% CI]	p
Change rate of alpha/beta bands	.621[.010, .9013]	.031*	.599[.050, .903]	.040*	.672[.085, .926]	.017*

N = 21, all correlations assessed two-tailed using a 2000 sample bootstrap with bias corrected accelerated confidence intervals.

4 Discussion

The aim of the study is to establish a connection between subjective perceptions of people's emotional reactions and objective physiological parameters, so that our findings can provide some preliminary insights into helping them easily learn about their own subjective psychological state.

Plenty researches on VR has come down to a conclusion that if lacks interaction, VR is just tool which may not be better than the traditional experimental tools [31]. Therefore, in our research, the subjects explored and experienced the emotional scenes provided as possible as they could to maximize the effectiveness of VR. According to the dynamic theory of emotional response, the temporal feature of emotional responsiveness takes an important role in human adaptation [6, 32, 33]. This means in fact that after the emotional evocation, it does not disappear immediately, but takes time to recover. The recovery time required by individuals was different. For example, watching horror pictures to induce emotions, some individuals have a longer duration of fear, and the recovery of various physiological parameter is slower. However, some people will be able to recover very soon [34]. Thus, the first 10 s after the end of the VR scene were chosen as an indicator of the emotions individual experienced.

Of particular interest in this study was to find effective indicators in physiological signals that reflect individual emotional reactivity. The dimensions of emotional reactivity included intensity, sensitivity, and persistence. Therefore, the difference of EEG data between VR stimulus and baseline levels were took into consideration trying to discover an appropriate indicator to describe individual ability to experience emotional intensity. Some studies on exercise training and brain function status assessment found that the relative power spectrum ratio of two frequency bands in the EEG signals is a relatively higher signal-to-noise ratio parameter [35, 36]. Our results also suggest that the change rate of alpha/beta ratio from left hemisphere is a valid indicator that can be used to assess an individual's emotional reactivity. The possible reason is that the ratio can reflect mental state of a person to a certain extent. The higher alpha/beta ratio after stimulus indicates that the mind is more relaxed, the brain is more fatigued, and the ability to process information is weaker, so the change rate means the ability to process an emotional information. It is positively correlated with emotional sensitivity and intensity. With a high change rate, one can easily react to emotional stimuli. All our participants are mentally healthy people, thus there is no orientation for negative stimuli [37]. The left hemisphere is mainly responsible for the processing of positive emotions [38], that is why this indicator only appeared on the left side of the brain.

Emotional persistence is an important dimension of emotional reactivity, a major manifestation of the temporal dynamics of emotions, and an important aim to emotion regulation [39, 40]. Good emotional persistence is mainly reflected in the goal of increasing the duration of positive emotions or reducing the intensity of negative emotions. Studies have shown the lateralization of human brain when processing emotional information [41]. Our findings regarding emotional persistence were mainly based on the valence hypothesis assumes that the left brain of human beings is responsible for the processing of negative emotions, while the right brain is responsible for the processing of positive emotions. Therefore, individuals with strong emotion regulation ability response differently to multiple emotional stimuli between the left and right hemisphere of brain in the expression of emotional persistence. The different responses over the recovery phase of the two hemispheres was chosen to express the persistence of emotion for the main factors considered searching for physiological indicators of emotional persistence are the changes that individual responded to certain stimulus over a period of time. As the result, it shows a significantly positive correlation with the change rate of the alpha bands ratio between right and left hemisphere after positive emotion and the change rate of the beta bands ratio between right and left hemisphere after negative emotion. This is consistent with the frontal EEG lateralization theory that the left and right prefrontal cortex hemispheres process differently for multiple emotions. During the activated phase, the energy of the alpha bands will decrease, and beta bands will increase. Therefore, the degree of activation in right brain differs from the left brain when processing positive and negative emotions, and the change rates of EEG between these two hemispheres are opposite under both emotional stimuli.

For several reasons the results of this current study should be considered only the first step in establishing a connection between objective parameters and subjective perceptions. There are some factors have not been taken into consideration associated with emotional reactivity such as emotion regulation strategy and coping style. These psychological traits are very important for emotional processing. For the better interpretation of individual's emotional reactivity, these factors should be included in further research. In addition, the physiological signal is composed of many different types of signals. However, for the reason of limited time, only EEG signal was preliminary analyzed, and the other signals such as heart rate and skin electricity still needed processing in future study. Multimodal affective computing should be the ultimate goal in this direction. In order to be able to better understand the relationship between emotional reactivity and physiological signals, and to apply these results to a broader social space, data collected from special community as a comparative study is necessary. At present, we have cooperated with some hospital starting to collect data of patients with first-episode depression.

5   Conclusion

Our research attempts to find a balance between subjectivity and objectivity and establish a relationship between these two. The outcomes of this study are consistent with the relevant theoretical basis of brain science. The individual EEG lateralization under different emotional stimuli can reflect the emotional intensity and emotional sensitivity perceived by themselves. The multi-band relative power spectrum ratio of normal group under positive stimuli conditions reflects individual's ability to maintain positive emotions.

Acknowledgements. This work is supported by "Science and Technology Program of Guangzhou" (201704020043), and "Natural Science Foundation of Guangdong Province, China" (2018A030310407).

References

1. Gross, J.J.: Antecedent-and response-focused emotion regulation: divergent consequences for experience, expression, and physiology. J. Pers. Soc. Psychol. **74**(1), 224–237 (1998)
2. Davidson, R.J.: Anxiety and affective style: role of prefrontal cortex and amygdala. Biol. Psychiat. **51**(1), 68–80 (2002)
3. Rothbart, M.K., Ahadi, S.A., Evans, D.E.: Temperament and personality: origins and outcomes. J. Pers. Soc. Psychol. **78**(1), 122–135 (2000)
4. Larsen, R.J., Diener, E.: Affect intensity as an individual difference characteristic: a review. J. Res. Pers. **21**(1), 1–39 (1988)
5. Davidson, R.J.: Emotion and affective style: hemispheric substrates. Psychol. Sci. **3**(1), 39–43 (1992)
6. Davidson, R.J.: Affective style and affective disorders: perspectives from affective neuroscience. Cogn. Emot. **12**(3), 307–330 (1998)
7. Linehan, M.M.: Cognitive-behavioral treatment of borderline personality disorder. Curr. Psychiatry Rep. **6**(3), 225–231 (1993)
8. Nock, M.K., Wedig, M.M., Holmberg, E.B., Hooley, J.M.: The emotion reactivity scale: development, evaluation, and relation to self-injurious thoughts and behaviors. Behav. Ther. **39**(2), 107–116 (2008)
9. Evans, S.C., Blossom, J.B., Canter, K.S., Poppert-Cordts, K., Kanine, R., Garcia, A.: Self-reported emotion reactivity among early-adolescent girls: evidence for convergent and discriminant validity in an urban community sample. Behav. Ther. **47**, 299–311 (2016)
10. Simon, S.C., Greitemeyer, T.: The impact of immersion on the perception of pornography: a virtual reality study. Comput. Hum. Behav. **93**, 141–148 (2019)
11. Weidner, F., et al.: Comparing VR and non-VR driving simulations: an experimental user study. In: 2017 IEEE Virtual Reality (2017)
12. Naz, A., Kopper, R., Mcmahan, R.P.: Emotional qualities of VR space. In: 2017 IEEE Virtual Reality (2017)
13. Corrêa, C.G.: Evaluation of VR medical training applications under the focus of professionals of the health area. In: ACM Symposium on Applied Computing (2009)
14. Picard, R.W.: Affective Computing. MIT Press, Cambridge (1997)
15. Radenkovic, MC.: Machine learning approaches in detecting the depression from resting-state electroencephalogram (EEG): a review study (2019)
16. Coan, J.A., Allen, J.J.B.: Frontal EEG asymmetry as a moderator and mediator of emotion. Biol. Psychol. **67**(1–2), 7–50 (2004)
17. Lin, Y.P., Wang, C.H., Jung, T.P., Wu, T.L., Chen, J.H.: EEG-based emotion recognition in music listening. IEEE Trans. Biomed. Eng. **57**(7), 1798–1806 (2010)
18. Liang, G.D., Li, Y.X., Liao, D., Hu, H.C., Zhang, Y.Y., Xu, X.M.: The relationship between EEG and depression under induced emotions using VR scenes. In: IMBioC (2019)
19. Rong, W.T.: Research on the brain cognition rule and EEG analysis method for image. Dissertation, Harbin Institute of Technology (2018)
20. Liao, D., et al.: Arousal evaluation of VR affective scenes based on HR and SAM. In: IMBioC (2019)
21. Li, G.: Research on the dynamics of emotional response and its influencing factors. Dissertation, Capital Normal University (2008)
22. Fingelkurts, A.A., Fingelkurts, A.A.: Timing in cognition and EEG brain dynamics: discreteness versus continuity. Cogn. Process. **7**(3), 135–162 (2006)
23. Lannoy, S., Heeren, A., Rochat, L., Rossignol, M., Van der Linden, M., Billieux, J.: Is there an all-embracing construct of emotion reactivity? Adaptation and validation of the emotion

reactivity scale among a French-speaking community sample. Compr. Psychiatry **55**(8), 1960–1967 (2014)

24. Shapero, B.G., Abramson, L.Y., Alloy, L.B.: Emotional reactivity and internalizing symptoms: moderating role of emotion regulation. Cogn. Ther. Res. **40**(3), 328–340 (2016)
25. Laurence, C., Smits, D., Bijttebier, P.: The Dutch version of the emotion reactivity scale (2014)
26. Li, Y.: Reliability and validity of the emotion reactivity scale in Chinese undergraduates. Chin. J. Clin. Psychol. **26**(02), 59–62 (2018)
27. Jadhav, N., Manthalkar, R., Joshi, Y.: Electroencephalography-based emotion recognition using gray-level co-occurrence matrix features. In: Proceedings of International Conference on Computer Vision and Image Processing (2017)
28. Simon, S., Greitemeyer, T.: The impact of immersion on the perception of pornography: a virtual reality study. Comput. Hum. Behav. **93**, 141–148 (2018)
29. Ralph, M.: Deep immersion with Kasina—an exploration of meditation and concentration within virtual reality environments. In: Computing Conference, London (2017)
30. Quoidbach, J., Berry, E.V., Hansenne, M.: Positive emotion regulation and well-being: comparing the impact of eight savoring and dampening strategies. Pers. Individ. Differ. **49**(5), 368–373 (2010)
31. Weidner, F., Hoesch, A., Poeschl, S., Broll, W.: Comparing VR and non-VR driving simulations: an experimental user study. In: Virtual Reality (2017)
32. Reich, J.W., Zautra, A.J., Davis, M.: Dimensions of affect relationships: models and their integrative implications. Rev. Gen. Psychol. **7**(1), 66–83 (2003)
33. Faith, M., Thayer, J.F.: A dynamical systems interpretation of a dimensional model of emotion. Scand. J. Psychol. **42**(2), 121–133 (2010)
34. Davidson, R.J.: Affective neuroscience and psychophysiology: toward a synthesis. Psychophysiology **40**(5), 655 (2003)
35. Xin, L., Xiaoying, Q., Yanxiu, T., Xiaoqi, S., Mengdi, F., Erjuan, C.: Application of the feature extraction based on combination of permutation entropy and multi-fractal index to emotion recognition. High Technol. Lett. **26**(7), 617–624 (2016)
36. Zhang, L.: Effects of beta band binaural beats on brain physiological status. Manned Spacefl. **22**(2), 254–261 (2016)
37. Van Bockstaele, B., et al.: The effects of attentional bias modification on emotion regulation. J. Behav. Ther. Exp. Psychiatry **62**, 38–48 (2018)
38. Lu, J.X.: Research on the emotional priming effect in the two hemispheres of the brain. Dissertation, Nanjing Normal University (2011)
39. Thompson, R.A.: Emotion and self-regulation. In: Nebraska Symposium on Motivation, vol. 36, p. 367 (1988)
40. Thompson, R.A.: Emotion regulation: a theme in search of definition. Monogr. Soc. Res. Child Dev. **59**(2-3), 25–52 (2010)
41. Zhang, J., Li, Y.S., Zhou, R.L.: The lateralization of the human brain to the processing of high and low arousal emotion pictures: research from event-related potentials. Chinese Psychological Society (2007)

Relevance of Common Spatial Patterns Ranked by Kernel PCA in Motor Imagery Classification

L. F. Velasquez-Martinez[1]([✉]), D. Luna-Naranjo[1], D. Cárdenas-Peña[2],
C. D. Acosta-Medina[1], G. A. Castaño[3], and G. Castellanos-Dominguez[1]

[1] Signal Processing and Recognition Group, Universidad Nacional de Colombia,
Bogotá, Colombia
`lfvelasquezm@unal.edu.co`
[2] Automatics Research Group, Universidad Tecnologica de Pereira, Pereira, Colombia
[3] Grupo de Trabajo Académico Cultura de la Calidad en la Educación,
Universidad Nacional de Colombia, Bogotá, Colombia

Abstract. Motor Imagery handles the brain activity patterns of motor action without explicit movements. For extracting the discriminating features, Common Spatial Patterns are the most widely used algorithm that is very sensitive to artifacts and prone to overfitting. Here, we develop a metric to assess the relevance of Common Spatial Patterns using a mapping through Kernel Principal Component Analysis with the benefit of improved interpretation that allows evaluating the zones, which contribute the most to the motor imagery classification accuracy. Validation is carried out on a real-world database, appraising two labels of Motor Imagery activity. From the obtained results, we prove that the developed approach allows the performance enhancement, at the time, the relevant set decreases the number of channels to feed the classifier, and thus reducing the computational cost.

Keywords: Common spatial patterns · KPCA · Motor imagery

1 Introduction

Brain-Computer Interface (BCI) systems rely on a cognitive neuroscience paradigm termed as Motor Imagery (MI). MI has particular applicability to neurophysical regulation and rehabilitation, games and entertainment, and sports training, among others. In education scenarios, for which the Media and Information Literacy methodology proposed by the UNESCO covers several competencies that are vital for people to be effectively engaged in all aspects of human development. The MI paradigm handles the brain activity patterns of the imagination of a motor action without an explicit movement taking into account that MI induces an Event-related Desynchronization/Synchronization over the corresponding spatial (channels) area over the sensorimotor cortex [3,6]. Nowadays, there are many challenges in rehabilitation in the development of systems

© Springer Nature Switzerland AG 2019
P. Liang et al. (Eds.): BI 2019, LNAI 11976, pp. 13–20, 2019.
https://doi.org/10.1007/978-3-030-37078-7_2

that enhance communication between the brain and the exterior environment. Conventional BCI have been widely used to assist disabled people to re-establish their capabilities of environmental control decoding the brain activity [13], being the electroencephalography (EEG) signal the most commonly used acquisition technique due to its implementation easiness and noninvasive implementation [9].

For extracting the discriminative spatial patterns, the algorithm of Common spatial patterns (CSP) is employed, contrasting the feature sets by the power measured for each MI class. Thus, the CSP method constructs spatial filters that maximize the variance of a specific class, but it simultaneously minimizes the variance of the remaining classes. However, since EEG is highly non-stationary, CSP is known to be very sensitive to artifacts and prone to overfitting. In practice, there is a need to find attribute subset preserving, as much as possible, input data variability to allow identifying the most discriminant information. Due to most of the relevance methods rely on the power estimates to extract the discriminative features, however, the presence of CSP sets with low amplitudes may result in relevant characteristics estimated very poor, introducing redundancy or even misleading the classifier. Besides, most of these feature selection methods are computationally expensive (mainly, heuristic approaches).

In this work, we develop a metric to assess the relevance of Common Spatial Patterns using a mapping through Kernel PCA with the benefit of improved interpretation that allows evaluating the zones, which contribute the most to the motor imagery classification accuracy. Validation is carried out on a real-world database, appraising two labels of BCI activity. From the obtained results, we prove that the developed approach allows the performance enhancement, at the time, the relevant set decreases the number of channels to feed the classifier, but reducing the computational cost.

2   Methods

Common Spatial Patterns as an Uncorrelated Space: Common spatial patterns (CSP) is a supervised feature extraction approach devoted to discriminating between a set of N band-pass filtered time series $\mathcal{X} = \{X_n^l \subset \mathbb{R}^{C \times T} : n \in N \subset \mathbb{N}\}$, where $l \in \{-, +\}$ are the labels, and $C \subset \mathbb{N}$ are the number of channels recorded along $T \subset \mathbb{N}$ time samples. CSP approach linearly maps the input EEG data from the channel space to a latent source space through the relationship: $Y = WX$, where $W \subset \mathbb{R}^{C' \times C}$ and $Y \subset \mathbb{R}^{C' \times T}$ are the linear mapping to $C' \leq C$ sources of the latent space, employing the class covariance $S^l \subset \mathbb{R}^{C' \times C'}$ computed as:

$$S_n^l = \mathbb{E}\left\{Y_n Y_n^\top\right\} = \mathbb{E}\left\{W X_n X_n^\top W^\top\right\},$$
$$= W \Sigma^l W^\top \tag{1}$$

being $\Sigma^l = \mathbb{E}\left\{X_n X_n^\top : \forall n \in N\right\}$ the normalized covariance of class l. Notation $\mathbb{E}\{\cdot\}$ stands for expectation operator.

For each row of W, denoted as $w_{c'} \subset \mathbb{R}^C$, CSP maximizes class covariance of the latent space by solving the optimization problem as below:

$$\max_{\boldsymbol{w}_{c'}} \boldsymbol{w}_{c'} \boldsymbol{\Sigma}^l \boldsymbol{w}_{c'}^\top \quad \text{s.t.} \quad \boldsymbol{w}_{c'} \left(\boldsymbol{\Sigma}^+ + \boldsymbol{\Sigma}^- \right) \boldsymbol{w}_{c'}^\top = 1$$

Therefore, signals in the CSP latent space are uncorrelated and their components with eigenvalues close to one or zero discriminate signals, according their variance.

Component-wise relevance analysis using kernel PCA: With the aim of associating between all trial components, each matrix of the provided variance set of uncorrelated CSP space $\mathcal{S} = \{\boldsymbol{S}_n : n \in N\}$ is further reorganized as $\boldsymbol{Z} = [\boldsymbol{z}_n^\top : n \in N]$, with $\boldsymbol{Z} \subset \mathbb{R}^{N \times C' \times C'}$, where $\boldsymbol{z} \subset \mathbb{R}^Q$ is the vectorized form of $\boldsymbol{S}_n$, and $Q = C' \times C'$. That is, $\boldsymbol{z} = [s_n^{(i,j)} : \forall i, j \in C']$

To perform Kernel Principal Component Analysis (KPCA), each column $\boldsymbol{z}_q$ is nonlinearly mapped to a higher dimensional space $\mathcal{F}$ through $\phi : \mathbb{R}^N \mapsto \mathcal{F}$ and PCA is performed in $\mathcal{F}$. The map ϕ is induced by a kernel function $k(\boldsymbol{z}_q, \boldsymbol{z}_{q'})$ that allows evaluating inner products in $\mathcal{F}$ as [12]:

$$k(\boldsymbol{z}_q, \boldsymbol{z}_{q'}) = \phi(\boldsymbol{z}_q)^\top \phi(\boldsymbol{z}_{q'}), \text{ with } \boldsymbol{z}_q, \boldsymbol{z}_{q'} \subset \mathbb{R}^N \tag{2}$$

Consequently, the kernel matrix $\boldsymbol{K} \in \mathbb{R}^{Q \times Q}$ holds the inner products of between columns of $\boldsymbol{Z}$ mapped onto $\mathcal{F}$. Such a kernel matrix which encodes the variability among time instants and CSP components. Aiming to quantify the variability of $\boldsymbol{K}$, the singular value decomposition is employed that extracts orthogonal eigenvectors $\boldsymbol{u}_m \subset \mathbb{R}^Q$ and non-negative eigenvalues $\lambda_m \geq 0$, so that the following expression holds:

$$\boldsymbol{K} = \boldsymbol{U} \boldsymbol{\Lambda} \boldsymbol{U}^\top \tag{3}$$
$$\boldsymbol{\Lambda} = \text{diag}(\lambda_1, \lambda_2, \dots, \lambda_Q)$$
$$\boldsymbol{U} = [\boldsymbol{u}_1, \boldsymbol{u}_2, \dots, \boldsymbol{u}_Q]$$

Taking advantage of above decomposition, we introduce the relevance measure in terms of the encoded variability as:

$$\rho = \sum_{m \in M} |\lambda_m \boldsymbol{u}_m|, \quad M \leq Q \tag{4}$$

where M is a free parameter that allows controlling the amount of variability considered to compute the relevance, and $\rho \in \mathbb{R}^Q$ is a non-negative relevance vector, where the larger the value - the more relevant the column $\boldsymbol{z}_q$.

In the attempt to identify the relevance of a pattern, ρ is a matrix reshaped in such a way that its elements represent a symmetric matrix of CSP patterns $\rho = [\rho_{c'd'} : c', d' \in C']$, with $\rho \subset C' \times C'$. Then, the relevance value for c'-th pattern is the summation over ρ columns as below:

$$\bar{\rho}_c = \mathbb{E} \{\rho_{c'd'} : \forall d' \in Q\}, \quad \bar{\rho}_{c'} \subset \mathbb{R}^+ \tag{5}$$

As a result, the relevance vector retains the variability of the input data in an uncorrelated space, so identifying the patterns contributing the most to discriminate two classes $(-, +)$, according to its encoded variability.

3   Experimental Set-Up

Database: We tested the proposed approach on the publicly available BCI competition IV dataset IIa[1]. This dataset includes a collation of EEG signals recorded from 22 electrodes. Nine subjects were instructed to perform four types of motor imagery activities (left hand, right hand, both feet, and tongue). Each subject performed two sessions on different days. Each session carried 6 runs with 48 trials (i.e., 12 trials per class), obtaining a total of 288 trials per session. Therefore, we conducted a bi-class classification task by selecting the motor imagery classes: left and right hand. Additionally, we removed from our analysis the trials marked with artifacts.

Training and Validation: The proposed approach aims at using KPCA for selecting the components that most contribute to the MI classification task. This approach appraises the following stages (see Fig. 1): (*i*) Preprocessing and CSP-based feature extraction, (*ii*) Component-wise analysis using KPCA, (*iii*) Extraction of relevant feature sets and bi-task classification, using linear discriminant analysis.

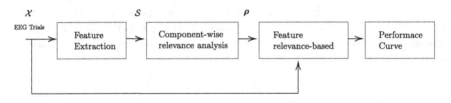

Fig. 1. Scheme illustrating the stages of proposed methodology.

In the preprocessing stage, all recordings were filtered employing a fifth-order Butterworth filter between 8 and 30 Hz in order to select the μ and β bands that are commonly linked to brain MI activity [5]. Furthermore, we selected the interval of 2.5–4.5 s considered as the MI activity interval [11]. In the feature extraction stage, we perform the CSP analysis using all components for W.

In the component relevance stage, we map the estimated components into a high-dimensional space, employing a kernel function with a free parameter corresponding to the bandwidth σ, which is tuned maximizing the variability of information potential [4]. Therefore, in the next stage, the assessed component relevances are used to compute the CSP-based features. Lastly, we obtain for each subject, the performance curves by adding relevant components using Linear Discriminant Analysis (LDA), using a $5\times$ fold validation.

---

[1] http://www.bbci.de/competition/iv/.

4   Results and Discussion

Figure 2 shows the MI discrimination accuracy presented as a function of the number of selected components according to the component relevance stage. As seen, when adding components ranked by the proposed component-wise relevance analysis, classification performance varies between subjects. For subjects $S8$ and $S9$, the accuracy increases to a particular value around 15 components, indicating that attaching more components is not adding valuable information to the classifier. Despite the notable variability, subjects $S1$ and $S7$ reach a higher performance using more components. For subjects labeled as $S2$, $S4$, $S5$, and $S6$, the learning curve exhibits large oscillations by adding components, maintaining low classification accuracy. Namely, by attaching more information does not represent a change significantly in the performance. These findings suggest the following: Due to the estimated components refer to the spatial filters that are used to project the EEG data getting signals that are optimally discriminate for left and right class with regard to the variance, the low performance infers that EEG channels from which we extracted components do not contain enough for discriminating the activity. In this regard, the components estimation CSP-based is affected by the signal-to-noise and non-stationary behavior of EEG data that can affect the class covariance matrices estimation, limiting the component or spatial-filter computation and yielding suboptimal results [2]. Besides, some subjects do not produce brain signals confidently in MI commonly associated with BCI-illiterate. Some works report between-subject variability in brain MI networks, suggesting that a BCI-illiterate subject has a less well-developed network for motor skills, which influence BCI performance [1].

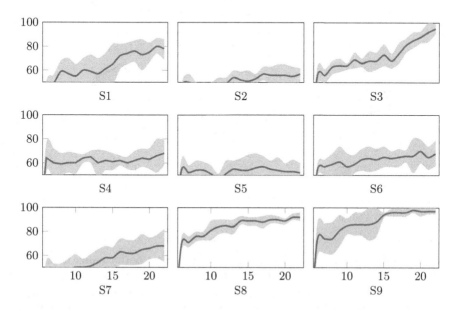

Fig. 2. Classification performance by adding channels according to component-wise relevance analysis.

Figure 3 displays the most relevant component selected by the proposed method, showing that all the relevant components exhibit discriminant information over the primary motor cortex, the motor area, and the primary somatosensory area; all these regions linked directly to MI tasks (see channels FC3 to C4) [14]. As suggested in [8], there are contributions from channels at the Parietal cortex commonly associated with the motor activity (see channels CP1 to CP4).

Overall, there is a relationship between performance and the channels that contribute to the most relevant component. For the subjects that have the highest accuracy (S9, S3, and S8), the channels over the motor cortex contribute the most. In turn, by joining the contribution of electrodes far from the sensorimotor zone, the classification accuracy tends to drop.

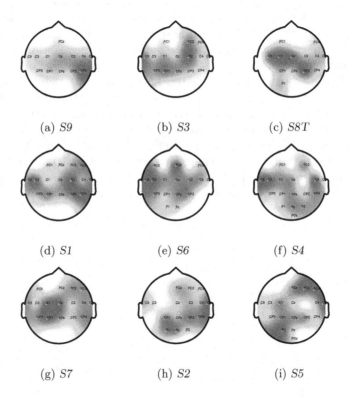

(a) S9   (b) S3   (c) S8T

(d) S1   (e) S6   (f) S4

(g) S7   (h) S2   (i) S5

Fig. 3. Estimated relevant components by subject. The figures were ranged by accuracy from the highest (Subject S8) to the lowest (Subject S5).

For the sake of comparison, Table 1 shows the subject performance contrasted against three benchmark methods that test the same database. As seen, the mean accuracy value provided by our proposed method outperforms the compared methods, also having comparable confidence. Of note, either comparative approach ([10] nor [7]) does not present any interpretation.

Table 1. Achieved classification results by subject. (average accuracy and standard deviation [%]).

Subject	Elasuty et al. [7]	Liang et al. [10]	Our approach
S1	76	62.13	**80.45 ± 6.5**
S2	45	**67.86**	57.38 ± 5.8
S3	92	75.71	**94.84 ± 4.23**
S4	**77**	72.14	68.21 ± 10.66
S5	64	**67.46**	57.35 ± 8.63
S6	**70**	66.67	69.88 ± 9.92
S7	59	**71.43**	68.49 ± 13.34
S8	**98**	78.57	92.47 ± 3.6
S9	80	70.00	**98.26 ± 2.38**
Mean	73.44	70.22	**76.37 ± 7.25**

As a concluding remark, we develop a metric to assess the relevance of Common Spatial Patterns using a mapping through Kernel PCA with the benefit of improved interpretation that allows evaluating the zones, which contribute the most to the motor imagery classification accuracy. In addition to the performance enhancement, the relevant set decreases the number of channels, reducing the computational cost. For future work, the authors plan to enhance the metric using more elaborate mappings. Validation of databases having a more considerable amount of subjects is also to be further considered.

Acknowledgments. This research was supported by Doctorados Nacionales, conv.727 funded by COLCIENCIAS, and the research project "Programa reconstrucción del tejido social en zonas de pos-conflicto en Colombia del proyecto Fortalecimiento docente desde la alfabetización mediática Informacional y la CTel, como estrategia didáctico-pedagógica y soporte para la recuperación de la confianza del tejido social afectado por el conflicto, Código SIGP 58950" funded by "Fondo Nacional de Financiamiento para la Ciencia, la Tecnología y la Innovación, Fondo Francisco José de Caldas con contrato No. 213 – 2018 con Código 58960".

References

1. Ahn, M., Jun, S.C.: Performance variation in motor imagery brain-computer interface: a brief review. J. Neurosci. Methods **243**, 103–110 (2015)
2. Alimardani, F., Boostani, R., Blankertz, B.: Weighted spatial based geometric scheme as an efficient algorithm for analyzing single-trial EEGS to improve cue-based BCI classification. Neural Netw. **92**, 69–76 (2017)
3. Allison, B.Z., Wolpaw, E.W., Wolpaw, J.R.: Brain-computer interface systems: progress and prospects. Exp. Rev. Med. Devices **4**(4), 463–474 (2007)
4. Álvarez-Meza, A.M., Cárdenas-Peña, D., Castellanos-Dominguez, G.: Unsupervised kernel function building using maximization of information potential variability. In: Bayro-Corrochano, E., Hancock, E. (eds.) CIARP 2014. LNCS, vol.

8827, pp. 335–342. Springer, Cham (2014). https://doi.org/10.1007/978-3-319-12568-8_41

5. Álvarez-Meza, A.M., Velásquez-Martínez, L.F., Castellanos-Dominguez, G.: Time-series discrimination using feature relevance analysis in motor imagery classification. Neurocomputing **151**, 122–129 (2015)

6. Blankertz, B., Tomioka, R., Lemm, S., Kawanabe, M., Muller, K.R.: Optimizing spatial filters for robust EEG single-trial analysis. IEEE Signal Process. Mag. **25**(1), 41–56 (2007)

7. Elasuty, B., Eldawlatly, S.: Dynamic bayesian networks for EEG motor imagery feature extraction. In: 2015 7th International IEEE/EMBS Conference on Neural Engineering (NER), pp. 170–173. IEEE (2015)

8. Hanakawa, T., Dimyan, M.A., Hallett, M.: Motor planning, imagery, and execution in the distributed motor network: a time-course study with functional MRI. Cereb. Cortex **18**(12), 2775–2788 (2008)

9. He, B., et al.: Electrophysiological brain connectivity: theory and implementation. IEEE Trans. Biomed. Eng. **66**(7), 2115–2137 (2019)

10. Liang, S., Choi, K.S., Qin, J., Wang, Q., Pang, W.M., Heng, P.A.: Discrimination of motor imagery tasks via information flow pattern of brain connectivity. Technol. Health Care **24**(s2), S795–S801 (2016)

11. Miao, M., Zeng, H., Wang, A., Zhao, C., Liu, F.: Discriminative spatial-frequency-temporal feature extraction and classification of motor imagery EEG: an sparse regression and weighted naïve bayesian classifier-based approach. J. Neurosci. methods **278**, 13–24 (2017)

12. Schölkopf, B., Smola, A.J., Bach, F., et al.: Learning with Kernels: Support Vector Machines, Regularization, Optimization, and Beyond. MIT press, Cambridge (2002)

13. Zhang, Y., Zhou, G., Jin, J., Zhang, Y., Wang, X., Cichocki, A.: Sparse bayesian multiway canonical correlation analysis for EEG pattern recognition. Neurocomputing **225**, 103–110 (2017)

14. Zich, C., Debener, S., Kranczioch, C., Bleichner, M.G., Gutberlet, I., De Vos, M.: Real-time EEG feedback during simultaneous EEG-fMRI identifies the cortical signature of motor imagery. Neuroimage **114**, 438–447 (2015)

Subject-Oriented Dynamic Characterization of Motor Imagery Tasks Using Complexity Analysis

L. F. Velasquez-Martinez$^{(\boxtimes)}$, F. Arteaga, and G. Castellanos-Dominguez

Signal Processing and Recognition Group, Universidad Nacional de Colombia,
Manizales, Colombia
Ifvelasquezma@gmail.com

Abstract. Motor Imagery is widely used applications, for which the analysis of underlying mechanisms is mostly focused on classification tasks. Relying on the Fractional Permutation Entropy (FPE), we propose a complexity-based analysis of individual brain dynamics extracted from motor imagery tasks. Due to conventional Common Spatial Pattern (CSP) filtering is affected by the high variability across trials, we perform the study of different stages of feature extraction, including raw data representation, conventional CSP, and high-level dynamics of the filter-banked CSP feature sets. Obtained results on a real-world application prove that promote the higher separation between patients extracting higher-level dynamics from features.

Keywords: Permutation entropy · Motor imagery · Bag-of-patterns

1 Introduction

As regards functional changes, brain neural activity is a complex multi-dimensional system, having a wide range of nonlinear and non-stationary dynamics that are interacting in response to a given stimulus by synchronization of oscillatory activities. In this regard, brain response could be useful in the development of Media and Information literacy (MIL) applications since it would allow to estimate the ability of a certain activator or stimulus to trigger a cognitive task [7]. Aiming to quantify the complexity of electroencephalographic signals (EEG), Entropy estimates are widely used in analysis of Motor-Imagery Systems, including Fuzzy Approximate Entropy [8], sample entropy [9], and, lastly, Permutation Entropy (PE) that is based on comparing neighboring values of each point to be mapped onto ordinal patterns [13]. However, an Entropy value must have adequate generality properties, being suitable for chaotic, regular, noisy, or real-world time series. On the other hand, the common spatial patterns (CSP) technique is a well-established approach to the spatial EEG filtering in MI applications, selecting those features that are maximally informative about the class labels. Due to CSP is affected by the high variability across trials

© Springer Nature Switzerland AG 2019
P. Liang et al. (Eds.): BI 2019, LNAI 11976, pp. 21–28, 2019.
https://doi.org/10.1007/978-3-030-37078-7_3

and the time-window for piecewise analysis [2], the CSP-extracted dynamics are weak and very noise-like.

This work discusses a complexity-based analysis of individual brain dynamics extracted from motor imagery tasks. We perform the study at different stages of feature extraction, including raw data representation, conventional Common Spatial filtering, and high-level dynamics of the filter-banked CSP feature sets.

2   Methods

MI dataset and preprocessing: The proposed spatio-spectral analysis is evaluated on the public collection of EEG signals recorded using a 22-electrode montage from nine subjects. Each subject performs two sessions on different days with four motor imagery tasks, namely left hand, right hand, both feet, and tongue on two sessions obtaining 144 trials for each class (BCI competition IV dataset IIa[1]). Being the most used MI paradigm in the literature, this work considers the binary MI task (namely, left and right movement). Therefore, we obtain a set $\{x_n^c \subset X \in \mathbb{R}^T : n \in N, c \in C\}$ that holds N acquired EEG recordings of length $T \in \mathbb{R}^+$. Also, $C \in \mathbb{N}$ is the number of channels to be further bandpass filtered, adjusting two main parameters for each subject: elemental bandwidth $B \subset F$ and their band overlapping $\delta_B \subset B$. Therefore, the following set of bandpass-filtered EEG data is obtained: $\{\tilde{x}_{n,b}^c \subset \tilde{X}_b : b \in B\}$. We use N_f band-pass filters with a bandwidth of $B = 4\,\text{Hz}$ as to cover the whole frequency EEG band $F \in \mathbb{R}^+$, ranging from 4 to 40 Hz and fixing the overlap between each other at $\delta_B = 2\,\text{Hz}$.

Bag-of-patterns representations using Common Spatial filtering: CSP finds a spatial filter matrix $W_b \in \mathbb{R}^{C \times 2K}$ to linearly map the bandpass-filtered EEG data $\tilde{X}_b \in \mathbb{R}^{T \times C}$ onto a space $\tilde{\tilde{X}}_b = W_b^\top \tilde{X}_b$, so that variance of the mapped signal is maximized for one class while the variance of another class is minimized [3]. In practice, a set of spatial filters $W_b^* = [w_{b,1}^* \dots w_{b,2K}^*]$ is obtained by collecting eigenvectors that correspond to the K largest and smallest eigenvalues of the generalized eigenvalue problem. Therefore, the CSP feature vector that accounts for the bandpass filtered components is formed as $\xi_b(\tilde{\tilde{X}}_b) = \log(\text{diag}\{\text{var}\{\tilde{\tilde{X}}_b\}\})$, being $\text{var}\{\cdot\}$ the variance operator and $\xi_b \in {}^{2K}$. Additionally, we repeated the same procedure for different N_τ overlapped time windows ($d_n = [\xi_b, \cdots, \xi_{N_f}]$ by the $n-$time window). Lastly, we built a feature matrix $D \in \mathbb{R}^{N \times 2K N_f N_\tau}$.

Each column vector d_i is an instance, and in turn, we assemble a bag using the whole column vector set: $B = [d_1^\top, \dots, d_{N_\tau}^\top]^\top$ with $B \in \mathbb{R}^{N_\tau \times 2M N_f}$. Therefore, a set of N_b training bags is denoted as $\mathcal{B} = \{B_i : \forall i \in [1, N_b]\}$, where each i-th bag holds N_i instance arrays, $B_i \in \mathbb{R}^{N_i \times Q}$, containing the instance vectors $b_{ij} \in \mathbb{R}^Q$, $j \in [1, N_i]$, where Q is the instance vector dimension. Each bag feature vector is the conditional probability that a z instance belongs to i-bag, $p_i \in [0, 1]^Z$ [5]:

$$P(b^z|B_i) \propto s(B_i, b^z) = \max_{\forall j \in N_i} \exp\left(-\|b_{ij} - b^z\|^2 / \sigma^2\right) \qquad (1)$$

---

[1] http://www.bbci.de/competition/iv/.

where $s(\boldsymbol{B}_i, \boldsymbol{b}^z)$, $s \in \mathbb{R}^+$, is the measure of similarity between the concept $\boldsymbol{b}^z$ and the bag $\boldsymbol{B}_i$, and $\sigma \in \mathbb{R}^+$ stands for the bandwidth of exponential square function. Consequently, the higher-level brain dynamics of motor imagery can be encoded within a Multiple-instance learning framework [4].

For feature selection in either case of representation $\boldsymbol{D}$, the LASSO sparse regression model is used [12]:

$$\boldsymbol{u}^* = \arg \min_{\boldsymbol{u}} \|\boldsymbol{D}\boldsymbol{u} - \boldsymbol{l}\|_2^2 + \nu \|\boldsymbol{u}\|_1 \qquad (2)$$

where $\boldsymbol{l} \in \mathbb{R}^{N_R}$ is a vector with the class labels, N_R is the trial set for available the two target variables, $\boldsymbol{u}$ is a sparse vector to be learned, $\nu \in \mathbb{R}^+$ is a positive regularization parameter for controlling the sparsity of $\boldsymbol{u}$.

Complexity analysis using Fractional Permutation Entropy (FPE):
Let $\boldsymbol{X} = [\boldsymbol{\xi}^\top(t) : t \in \{1, n - (m-1)\tau\}]$ be the phase space reconstruction matrix with embedding dimension m, holding a set of m-dimensional reconstruction vectors for each sample $\boldsymbol{\xi}(t) = [x(\zeta) : \zeta \in \{t, t + (m-1)\tau\}]$, where $\boldsymbol{x} = [x(t) : t \in T]$ is an one-dimensional time-series, from which ordinal patterns can be generated. Note that the transformed component is rearranged by ranking in ascending order the magnitude values as follows: $\zeta(t + (j_1 - 1)\tau) \leq \zeta(t + (j_2 - 1)\tau) \leq \ldots \leq \zeta(t + (j_m - 1)\tau)$, being $j_1, \ldots, j_m$ the column index where each element is located in the transformed component. A set of different ordinal patterns $\{\boldsymbol{s}^l = [j_1 j_2 \ldots j_m] : l \in m!\}$ can extracted from each row $\boldsymbol{\zeta}(t)$. So, the associated relative frequencies (i.e., the probability of each symbol sequence, p_j) calculated can be directly computed by the number of times the particular m-order sequence is found in the time series divided by the total amount of sequences. Lastly, the previous probability set is used to compute the permutation entropy of order m for the time series $\boldsymbol{x}$ as follows: $\theta_p(m) : \boldsymbol{x} \in \mathbb{R}^T \mapsto \mathbb{R}^+$, $\theta_p(m) = \sum_{j \in m!} p_j \ln p_j$. Hence, the smaller the PE, the more regular and more deterministic the time series. In other words, whenever $\theta_p(m)$ gets close to 1, the time series is likely to be more noisy. Here, we implement the FPE approach for PE given in [6].

3 Experimental Set-Up

For validation purposes, the subject-oriented dynamic characterization through the complexity analysis comprises three stages (see Algorithm 1): (*i*) Complexity evaluation of measured EEG time-series, (*ii*) Evaluation of CSP-based feature dynamics extracted from measured input EEG data, (*iii*) Complexity evaluation of high-level bi-tasks brain dynamics obtained from CSP-based representation.

In the beginning, aiming to make clear the contribution of each scalp EEG channel in terms of variability, the FPE calculation is performed, adjusting previously the $\alpha = 0.2$ parameter $(0 < \alpha \leq 0.5)$ to rule the local order structure of underlying extracted dynamics. The top row in Fig. 1 displays the FPE values of each EEG channel performed by each subject $\theta_m(c|l)$, showing that even that most of the channels have similar complexity, there are few electrodes with more

---

Algorithm 1. Validation algorithm of the performed subject-oriented dynamic characterization through complexity analysis using bi-tasks MI data.

Data: Given the C-montage scalp EEG data $\boldsymbol{X}_m$, subject set $m \in M$, trial set $n \in N$, label set $\lambda \in \{l, l'\}$.

for *each m and λ* **do**

    compute $f : \boldsymbol{X}_m^{(n)} \in \mathbb{R}^{C \times T} \mapsto \boldsymbol{\theta}_m^{(n)} \in \mathbb{R}^{N_C}. c \in C$

    Average across n: $\theta_m(c) = \mathbb{E}\left\{\theta_m^{(n)}(c) : \forall n \in N\right\}$

end

Compute $\bar{\theta}(c) = \mathbb{E}\{\theta_m(c) : \forall m \in M\}$, $\sigma_\theta(c) = \mathrm{std}(\theta_m(c) : \forall m \in M)$

Compute $\rho(m, m'|\lambda) = \langle\theta_m(c|l), \theta_{m'}(c|l)\rangle$, $\forall m \in M$.

Output: Evaluation of input measured EEG dynamics. See Fig. 1

Data: Given the input CSP features space – $\boldsymbol{D}_m$, $m \in M$, $\Delta f \in F$ $\lambda \in \{l, l'\}$.

for *each m and λ* **do**

    Compute $f : \boldsymbol{D}_m^{(n)} \in \mathbb{R}^{N \times 2KN_f N_\tau} \mapsto \boldsymbol{\theta}_m^{(n)} \in \mathbb{R}^{N_f}.$

    Average across n: $\theta(c) = \mathbb{E}\left\{\theta^{(n)}(f) : \forall n \in N_f\right\}$

end

Compute $\bar{\theta}^{(n)}(f) = \mathbb{E}\{\theta_m(c) : \forall m \in M\}$, $\sigma_\theta(c) = \mathrm{std}(\theta_m(c) : \forall m \in M)$

Compute $\rho(m, m'|\lambda) = \langle\theta_m(c|l), \theta_{m'}(c|l)\rangle$, $\forall m \in M$;

Output: Evaluation of input filter-banked CSP dynamics. See Fig. 2

---

information. It is worth noting that this situation holds similarly for all subjects, and regardless of the performed task (labels l, l') as shown by the mean value $\bar{\theta}^{(n)}(c|l)$ (painted in purple) averaged across the whole subject set.

Another aspect to remark is that each FPE estimate $\theta_m(c|l)$ has a different offset for each subject, meaning that the used measure may allow characterizing the differences of complexity between each patient reasonably. However, as a whole, brain dynamics between channels behave similarly. The bottom row in Fig. 1 presents the computed correlation index $\rho(m, m'|\lambda)$ as a pairwise similarity measure, showing a high similarity between subjects despite the MI task label and therefore being difficult the distinction of individual brain dynamics at this initial stage.

In the next stage, we perform a complexity evaluation of the CSP time-series extracted. The top row in Fig. 2 displays the FPE values $\theta_m(\Delta f|l)$ calculated across the spectral bandwidths, $\Delta f \in F$, and shows that the individual complexity trajectories gather closer to the mean value $\bar{\theta}^{(n)}(\Delta f|l')$. Note that we set-up $N_f = 17$ using five-order overlapped bandpass Butterworth filters and $K = 3$ for Common Spatial Filtering stage, as suggested in [14].

Generally speaking, the frequency band filtering is performed to improve robustness at a low signal-to-noise ratio, exploits the spectral relationship among the extracted MI features [10]. This fact can be evidenced by the values of accuracy shown in Table 1, and estimated by the approach detailed in [1] for time-series discrimination using feature relevance analysis in motor imagery

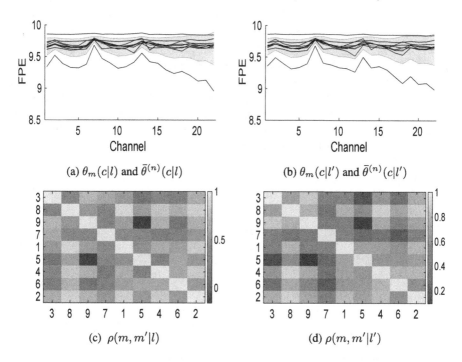

Fig. 1. FPE of measured EEG data (top row) and correlation index of complexity between subjects (bottom row). Left column - left hand task, right column - right hand.

classification. As seen, the mean accuracy value is above 80% with subjects reaching 94%, and implying competitive figures of discrimination between classes.

As regards individual discrimination, the filter-banked CSP features allow improving separation between subjects as seen in Table 1 that presents lower correlation index values than the ones assessed between subjects for the ones performed by the measured input EEG data. For the convenience of interpretation, the horizontal axis displays all subjects ranked in decreasing order of assessed accuracy. Therefore, along with the improvement in accuracy of bi-class MI tasks, filter-banked CSP features promote also the higher separation between patients.

Lastly, we performed the complexity analysis over the enhanced bag-of-patterns representation, or which instances are constructed by Filter-banked CSP feature sets extracted in accordance with Eq. (1). In this case, we rely on the LASSO fits (vector u), which are estimated by the feature selection task, and calculated at each time instant. Aiming to improve the confidence of assessed fits, different trajectories represent the estimation performed for several cross-validated folds as presented in Fig. 3a.

For evaluating the relationship between the subjects, however, it is more appropriate to consider the similarity based on the higher-level arrangements

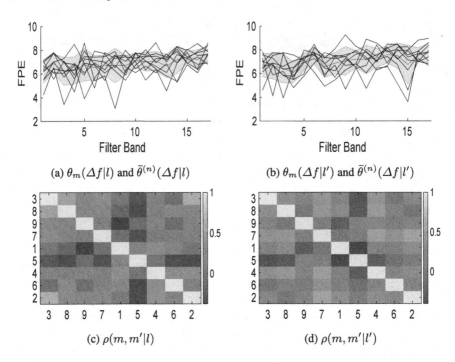

(a) $\theta_m(\Delta f|l)$ and $\bar{\theta}^{(n)}(\Delta f|l)$

(b) $\theta_m(\Delta f|l')$ and $\bar{\theta}^{(n)}(\Delta f|l')$

(c) $\rho(m, m'|l)$

(d) $\rho(m, m'|l')$

Fig. 2. Computed fractional PE of filter-banked CSP features within each bandwidth (top row) as well as the correlation index of PE between subjects (bottom row). Left column - left hand task, right column - right hand.

Table 1. Performed values of classification accuracy by finding of the time-varying features influencing the most on distinguishing bi-class MI tasks.

Subject	1	2	3	4	5	6	7	8	9
Accuracy(%)	89.64	67.71	94.40	73.78	83.17	71.88	90.60	92.79	91.48

in Eq. (1), aiming to capture more accurately the structural dynamics to be fed into a classifier and without having to deal with the raw data [11]. The results of similarity between subjects for the used bag-of-patterns representation are estimated across the fold time-series of LASSO fits u. Both plots Figs. 3b and c represent the values ranked in decreasing order by the assessed accuracy in Table 1 and by the computed FPE, respectively. In the former ranking case, there is a very high distinction (similarity close to 0) between the subjects that perform the bi-class MI task accurately. On the opposite, there is a high similarity between all subjects completing the worst. Likewise, the PFE achieves comparable results of similarity between the high-level brain dynamics extracted from MI bi-tasks.

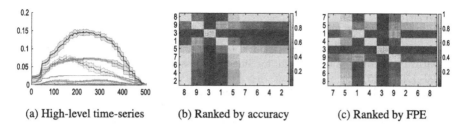

(a) High-level time-series   (b) Ranked by accuracy   (c) Ranked by FPE

Fig. 3. Subject-oriented complexity analysis performed over the high-level representation of brain neural dynamics.

Concluding Remark

Relying on the Fractional Permutation Entropy, we propose a complexity-based analysis of individual brain dynamics extracted from motor imagery tasks. We perform the study of different stages of feature extraction, including raw data representation, conventional Common Spatial filtering, and high-level dynamics of the filter-banked CSP feature sets. From the validation results obtained from the BCI competition IV dataset IIa, the following aspects are worth to be mentioned:

The use of complexity analysis of the brain neural activity of nonlinear and non-stationary dynamics, contributing to the distinction between individuals even if the EEG data is devoted for classification purposes. However, the representation of underlying brain dynamics plays an important role. Thus, it is almost impracticable to distinguish between individual brain dynamics using the EEG raw data. However, using conventional feature extraction techniques, like CSP, the similarity between subjects decreases. That is, along with the improvement in accuracy of bi-class MI tasks, filter-banked CSP features also promote the higher separation between patients.

For evaluating the relationship between the subjects, however, it is more appropriate to consider the similarity based on the higher-level dynamics. In particular, the bag-of-patterns representation a very high distinction (similarity close to 0) between the subjects, but only in the cases of performing the bi-class MI task accurately.

As future work, the authors plan to explore more robust measures of complexity, intending to deal with sensitivity under very noise measurements. Also, another bag-of- patterns representations are to be considered for extracting more accurately the individual brain dynamics.

Acknowledgments. This research was supported by Doctorados Nacionales, conv.727 funded by COLCIENCIAS, and the research project "Programa reconstrucción del tejido social en zonas de pos-conflicto en Colombia del proyecto Fortalecimiento docente desde la alfabetización mediática Informacional y la CTel, como estrategia didáctico-pedagógica y soporte para la recuperación de la confianza del tejido social afectado por el conflicto, Código SIGP 58950" funded by "Fondo Nacional de Financiamiento para la Ciencia, la Tecnología y la Innovación, Fondo Francisco José de Caldas con contrato No. 213 − 2018 con Código 58960".

References

1. Alvarez-Meza, A., Velasquez-Martinez, L., et al.: Time-series discrimination using feature relevance analysis in motor imagery classification. Neurocomputing **151**, 122–129 (2015)
2. Ang, K., Chin, Z., et al.: Filter bank common spatial pattern algorithm on BCI competition IV datasets 2a and 2b. Front. Neurosci. **6**, 39 (2012)
3. Blankertz, B., Tomioka, R., et al.: Optimizing spatial filters for robust eeg single-trial analysis. IEEE Signal Process. Mag. **25**(1), 41–56 (2007)
4. Caicedo-Acosta, J., Cárdenas-Peña, D., Collazos-Huertas, D., Padilla-Buritica, J.I., Castaño-Duque, G., Castellanos-Dominguez, G.: Multiple-instance lasso regularization via embedded instance selection for emotion recognition. In: Ferrández Vicente, J.M., Álvarez-Sánchez, J.R., de la Paz López, F., Toledo Moreo, J., Adeli, H. (eds.) IWINAC 2019. LNCS, vol. 11486, pp. 244–251. Springer, Cham (2019). https://doi.org/10.1007/978-3-030-19591-5_25
5. Chen, Y., Bi, J., et al.: MILES: multiple-instance learning via embedded instance selection. IEEE Trans. Pattern Anal. Mach. Intell. **28**(12), 1931–1947 (2006)
6. Chu, C., Wang, J., et al.: Complexity analysis of EEG in AD patients with fractional permutation entropy. In: 2018 37th Chinese Control Conference, pp. 4346–4350 (2018)
7. Frau-Meigs, D.: Media Education. Parents and Professionals. Unesco, A Kit for Teachers, Students (2007)
8. Hsu, W.-Y.: Assembling a multi-feature EEG classifier for left-right motor imagery data using wavelet-based fuzzy approximate entropy for improved accuracy. Int. J. Neural syst. **25**(8), 1550037 (2015)
9. Liu, Y., Huang, S., et al.: Novel motor imagery-based brain switch for patients with amyotrophic lateral sclerosis: a case study using two-channel electroencephalography. IEEE Consum. Electron. Mag. **8**(2), 72–77 (2019)
10. Miao, M., Wang, A., et al.: A spatial-frequency-temporal optimized feature sparse representation-based classification method for motor imagery EEG pattern recognition. Med. Biol. Eng. Comput. **55**(9), 1589–1603 (2017)
11. Passalis, N., Tsantekidis, A., et al.: Time-series classification using neural Bag-of-Features. In: 2017 25th European Signal Processing Conference (EUSIPCO), pp. 301–305 (2017)
12. Shin, Y., Lee, S., et al.: Sparse representation-based classification scheme for motor imagery-based brain-computer interface systems. J. Neural Eng. **9**(5), 056002 (2012)
13. Zanin, M., Gómez-Andrés, D., et al.: Characterizing normal and pathological gait through permutation entropy. Entropy **20**(1), 77 (2018)
14. Zhang, Y., Zhou, G., et al.: Optimizing spatial patterns with sparse filter bands for motor-imagery based brain-computer interface. J. Neurosci. Methods **255**, 85–91 (2015)

Modeling Individual Tacit Coordination Abilities

Dor Mizrahi, Ilan Laufer, and Inon Zuckerman$^{(\boxtimes)}$

Department of Industrial Engineering and Management,
Ariel University, Ariel, Israel
inonzu@ariel.ac.il

Abstract. Previous experiments in tacit coordination games hinted that some people are more successful in achieving coordination than others, although the variability in this ability has not yet been examined before. With that in mind, the overarching aim of our study is to model and describe the variability in human decision-making behavior in the context of tacit coordination games. To do so we first conducted a large-scale experiment to collect behavioral data, modeled the decision-making behavior, and characterize their observed variability. We then used the proposed model by predicting the individual coordination ability of a player based on its constructed strategic profile model and demonstrated that there is a direct and significant relationship between the player's model and its coordination ability. Understanding the differences in individual's tacit coordination abilities as well as their unique strategic profiles will allow us to better predict human's behavior in tacit coordination scenarios and consequently construct better algorithms for human-machine interactions.

Keywords: Tacit coordination games · Decision making · Cognitive modeling

1 Introduction

A *tacit coordination game* is one in which two individuals are rewarded for making the same choice from the same set of alternatives, and any form of communication between the players is not allowed or not possible [1]. Such problems have been formally modeled in game theory as games with multiple Nash equilibria solutions with equal values [2]. Nash equilibrium is a solution concept within game theory where each player plays his best response strategy. There are games where there are several points of Nash equilibrium, the possibility of multiple equilibria causes the outcome of the game to become less predictable - which led to the problems of coordination games. Since Schelling's seminal work in 1960 [3] many experiments have shown that people somehow manage to converge to a solution more effectively than what was predicted by the game-theoretical analysis. Apparently, for different reasons, some equilibria solutions appear more prominent than others for the players in the game. These solutions are denoted as *focal points*. In contrast, the game-theoretical framework often fails to

© Springer Nature Switzerland AG 2019
P. Liang et al. (Eds.): BI 2019, LNAI 11976, pp. 29–38, 2019.
https://doi.org/10.1007/978-3-030-37078-7_4

explain people decisions in such games [1], mostly because the problem of deciding between multiple Nash equilibria is one of games theory biggest challenges [4, 5].[1]

Another question that has been left unanswered is the degree of heterogeneity in peoples' ability to successfully coordinate in tacit coordination games. In other words, given a set of tacit coordination problems, it seems that some people manage to successfully coordinate most of their answers with the unknown partner, while others experience difficulties in doing so. This ability to succeed in tacit coordination tasks was hinted at by Bacharach in [1] and was named *"Schelling's competence"*.

The overarching aim of our study is to model and describe the variability in human decision-making behavior in tacit coordination games. To do so we conducted a large-scale tacit coordination experiment to collect behavioral data. We analyzed and modeled the participants decision-making behavior as a *strategic profile* and characterize the observed variability in these profiles. Next, we validated the proposed model by predicting the individual coordination ability of a player based on its constructed strategic profile. Understanding the differences in individual's tacit coordination abilities as well as their unique strategic profiles will allow us to better predict human's behavior in tacit coordination scenarios and consequently construct better algorithms for human-machine interactions.

2   Materials and Methods

2.1   Experimental Design

To test the coordination ability of players we used the "Assign Circles" [6, 7] tacit coordination game. The players were presented with 14 different "Assign Circles" game boards: 10 predefined games taken from [7] and in 4 additional games the number of circles ($\sim$ Uniform [1, 7]) and their placement were randomized for each player. In each of the games the players were asked to assign circles to squares with the aim of coordinating their assignment with an unknown player, who is presented with the same board (see Fig. 1 for a game example). That is, *a successful coordination is achieved when both players attached all the circles to the same squares*. In case of a successful coordination both players gain a point and the score is accumulated as the game progresses. Both players have no communication capability at all, and their results are only revealed after all the games have been completed.

The authors built upon [6, 7] to postulate that participants would utilize three main selection rules when playing the "Assign Circles" games, as follows: (1) *closeness* – assigning each circle to the closest available square; (2) *accession* – assigning circles which are close to one another to the same square; and (3) *equality* – evening out the number of circles that are assigned to each of the two available squares. We examined the variability in how these three rules were weighted and aggregated as part of the decision-making process of each individual participant. Importantly, we do not claim that these three rules represent an exhaustive list of rules from which a participant can

---

[1] Quoting Binmore ([2], p. 262), *"The equilibrium selection problem is perhaps the greatest challenge facing modern game theory"*.

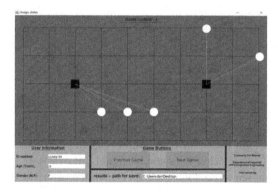

Fig. 1. "Assign Circles" application window

choose (see [7]). Rather, we merely suggest that these three prominent selection rules should provide enough variability in the individual strategic profiles.

As the first ten games have a fixed layout, we can analyze each one of them using the abovementioned selection rules and detect the expected solution by implementing each of the rules in each game. In this way, we may potentially predict the solutions by which the players will choose to establish a focal point if the strategic profile for each participant is consistent between games. For example, let us examine game from Fig. 1. In this game, first, it is easy to see that the *equality* rule is not applicable as there are 5 circles (any division of 2–3 or 3–2 will not satisfy the equality condition). Second, the *accession* rule solves the coordination problem exactly as is illustrated in the figure (we can visually detect two groups of circles). Lastly, using the *closeness* rule we expect the middle circle to be connected to the right-hand square instead of the one it is connected to in the figure. In a similar manner we can compute the solution attained by using each of the selection rules for each game instance.

The participants were 93 students that were enrolled in one of the courses on campus (49 of whom were female, mean age = ~ 23, SD = 1.97). The 93 subjects were randomly allocated to two consecutive sessions, with 48, and 45 participants in each of the sessions, respectively. All subjects were seated in one class and each student sat in front of a desktop monitor. Before the onset of the experiment, participants received an explanation regarding the overarching aim of the study, the experimental procedure, and the graphics of the application window. As participants were rewarded according to their performance on the "Assign Circles" task, participants also received an explanation regarding the incentive scheme used to allocate the rewards (in the form of different levels of course credit).

2.2 Measures Related to the Experiment

Coordination Index (CI). The CI measure proposed by Mehta et al. [7] is a statistical measure that allows determining the difficulty of coordination in a specific game. Specifically, the higher the CI, the easier it is to coordinate between two random players. Consider a coordination game with the label set $L = \{l_1, \ldots, l_n\}$, and any set of

N individual players, when each of them plays the game only once with an unknown anonymous partner. For each label l_j let m_j be the number of individuals who choose it, then the coordination index c is given by:

$$c = \frac{\sum_j m_j(m_j - 1)}{N(N - 1)} \tag{1}$$

This index measures the probability that two distinct individuals, chosen at random without replacement from the set of N individuals, choose the same label. It takes the value 1 if all individuals selected the same label and 0 if everyone selected a different label. If labels are chosen at random, the expected value of the index is $\frac{1}{n}$.

Individual Coordination Ability (ICA). To measure the coordination ability of each player, we assessed their ability to coordinate with all 92 other participants in the experiment rather than with a single random participant. For this purpose, we calculated the total number of games in which each player was able to coordinate his responses against the entire experimental population and normalized by the total number of games. It should be noted that the calculation was only carried out on the ten predefined games, but not on the four randomized games, since only the predefined games were kept constant across participants. ICA is formally defined as follows:

$$ICA(i) = \sum_{j=1|(j \neq i)}^{N} \sum_{k=1}^{t} \frac{CF(i,j,k)}{(N-1) * t} \tag{2}$$

Where i denotes the ith participant, N denotes the total number of participants, and t denotes the number of games in the experiments. The CF (coordination function) is defined as follows:

$$CF(i,j,k) = \begin{cases} 1; & \text{if players i and j chose the same label in game k} \\ 0; & \text{otherwise} \end{cases} \tag{3}$$

The ICA measure is not intended to be an absolute score, but rather it allows ranking the participants that completed the same set of tasks based on their ICA values as was the case in our study.

Strategy Rate (SR). The strategy profile is described by the weighted combination of the selection rules. To measure the frequency of choosing each of the abovementioned strategies by a single player, we first defined the Strategy Rate (SR). This measure reflects the probability of using a specific selection rule by a specific player based on the behavioral performance data. The SR defines the probability of using each of the three selection rules while playing the games. SR is defined as the probability that a specific player will choose to use a specific strategy in one of the Assign Circles games (i represents the ith player). The SR was calculated based on the behavioural performance data of each player in each game included in the Assign Circles game. We first define a Game Tag (GT) variable for each strategy, which can take one of three different values as follows:

$$GT(i,k)_{\text{strategy}} = \begin{cases} 1, & \text{if the strategy was available in the } k^{th} \text{ game and the } i^{th} \text{ player used it} \\ 0, & \text{if the strategy was not available in the } k^{th} \text{ game} \\ -1, & \text{if the strategy was available in the } k^{th} \text{ game and the } i^{th} \text{ player didn't use it} \end{cases}$$

$$(4)$$

The SR for each of the three game strategies can then be calculated as follows:

$$SR(i)_{\{\text{Acc,Equ,Clo}\}} = \frac{\sum_{k=1}^{14}\left[GT(i,k)_{\{\text{Acc,Equ,Clo}\}} = 1\right]}{\sum_{k=1}^{14}\left|GT(i,k)_{\{\text{Acc,Equ,Clo}\}}\right|} \tag{5}$$

Together, the strategic profile of each individual player is a vector composed of three elements, one for each of the selection rules ($\{\text{Acc}, \text{Equ}, \text{Clo}\}$) using Eq. 5.

3   Results and Discussion

Before moving to analyzing the results, we first need to demonstrate that a solution in each of the games was not randomly selected. In other words, that the players were motivated to coordinate with the other unknown partner. In Table 1 we compare the CI with the results of the CI that would have been obtained had all the subjects acted by picking their solution *randomly*, as is expected by game theory. We can see that all games had a significantly higher CI values than random picking (a higher score denotes a better ability to coordinate). We can also notice that the difficulty of coordination varies across different games, as reflected by the fluctuations in the CI in the different games. There are few easy games (e.g. #1, #2), few hard games (e.g. #3, #8) and few games with medium difficulty level (e.g. #4, #5).

Table 1. Comparison of CI values to random picking

Game number	1	2	3	4	5	6	7	8	9	10
CI	0.71	0.75	0.36	0.43	0.55	0.52	0.74	0.36	0.33	0.41
Random picking CI	1/4	1/16	1/8	1/16	1/8	1/16	½	¼	1/16	1/32

3.1   Characterizing the Variability in the Strategic Profiles

To characterize the variability in the coordination ability among different players, we have plotted the distribution of the data using a violin plot combined with boxplot, a histogram and a Gaussian mixture model (GMM) (Fig. 2A, B and C, respectively). We can see that the range of the ICA score is very diverse: the lowest score was 0.023 (average of 2% successful coordination) and the highest score 0.666 (average of 66% successful coordination). Also, we can see that the top players in our domain are somewhat homogenous group having an ICA value in the [0.55, 0.7] range. It can also be observed that the distribution is negatively skewed and that the lower 25% ICA scores are more spread out than the higher scores (Fig. 2A). In addition, the Dip test of unimodality

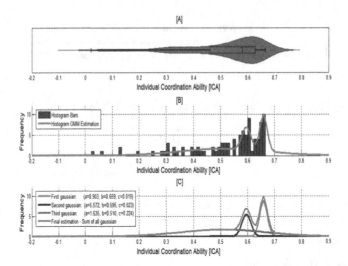

Fig. 2. The distribution of the CA scores. (A) Violin Plot with Boxplot (B) Histogram (C) Gaussian Mixture model estimation. The abscissa is the CA score. Note that the third Gaussian (green line) is obsqured by the general model because of its high variability. (Color figure online)

[8] indicated that the distribution is multimodal (p > 0.1, Dip = 0.025) while by estimating the distribution by a GMM with K Gaussians it was demonstrated that the entire distribution can be described by three Gaussians (k = 3 was determined by using Residual sum of squares). The mathematical formulation of the ICA mixture model, which is presented graphically in Fig. 2C, can be described using the following formula:

$$ICA(x) = 8.903 * e^{-\left(\frac{(x-0.659)}{0.019}\right)^2} + 5.572 * e^{-\left(\frac{(x-0.595)}{0.023}\right)^2} + 1.635 * e^{-\left(\frac{(x-0.510)}{0.224}\right)^2}$$

It is noteworthy that this division into k = 3 was also corroborated using the silhouette index [9] for K values at the range [1, 10]. The three Gaussians composing the distribution (Fig. 2C) comprise the third Gaussian with the lowest ICA scores which corresponds with the lower 25%, the second Gaussian which surrounds the median, and the third Gaussian which corresponds with the upper portion of the IQR as well as with the upper quartile (Fig. 2A and B).

3.2 Detection of Dominant SR Values in the Strategic Profile

As a preliminary step, before we analyze the relationship between ICA and the strategic profile, we first must identify the predominant strategy selected by each of the players. Table 2 displays the distribution parameters of each of the three SR indices:

Table 2. Strategy rate statistics

	$SR_{\text{Closeness}}$	SR_{Equality}	$SR_{\text{Accession}}$
Mean	0.6919	0.6685	0.3439
Standard deviation	0.2690	0.2624	0.1569
Median	0.75	0.7143	0.3636

It is clear from Table 2 that there were two leading selection rules: closeness, and equality. The median probability of using each of the two selection rules by a random player was over 70%, if they were applicable in the game. In contrast, the accession rule had a much lower probability of selection on average by a random player.

A one-way analysis of variance showed that the main effect of strategy rate was significant $F(2, 90) = 63.67$, $p < 0.001$. Post hoc analyses using Tukey's honestly significant difference criterion [10] indicated that the average mean value of the accession SR value was significantly lower than in the other two strategy SR values, closeness, and equality ($p < 0.001$) (the latter two reported p-values are in the order of 10^{-9}). As for the comparison results of the remaining two strategies, closeness and equality, we received no statistical significance with p-value of 0.78.

3.3 Modeling the Connection Between Strategic Profile and ICA

The previous section showed that the strategic profile contains two leading strategies (closeness and equality) and another secondary strategy (accession). This phenomenon may affect the performance of the predictive model due to unwanted dependencies between the independent variables (shown in the ANOVA analysis above). On the other hand, the omission of one of these variables will result in a great loss of information which will impair the performance of the predictive model. To deal with that problem we will use a dimension reduction technique to maximize the variance of the data in a smaller number of features, thereby reducing the dependencies between the variables. To do so we used the Principal Component Analysis (PCA) algorithm.

With PCA we find the new base vectors, named principal components and denoted by U in Fig. 3. The original data is then projected onto the new base vectors resulting in a new dataset with fewer features and with minimal dependence between them. Figure 3 presents the original data (based on the three SR values), with the corresponding principal components found by calculating the Eigenvectors of the covariance matrix of the data after preprocessing of mean and variance normalization:

To ensure that no critical data was lost during the process of dimensionality reduction, we calculated the retained variance in the process. The process was carried out by compressing the original data by reducing the dimensionality of the data. Next, we performed a reconstruction of the reduced data into the original dimension while comparing the reconstructed data to the original data vector before compression. The

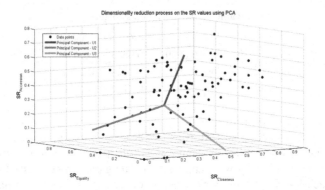

Fig. 3. Dimensionality reduction process on the SR values using PCA. All three SR points of each player (closeness, equality, and accession) are displayed within a 3D plane, together with the principal components obtained by the dimensionality reduction proces.

strategic profile of the ith player is denoted by $x^{(i)}$ vector, which contains the three SR values. The retained variance calculation is performed as follows:

$$retained\ variance = 1 - \frac{\frac{1}{m}\sum_{i=1}^{m}\left|\left|x^{(i)} - x_{approx}^{(i)}\right|\right|^2}{\frac{1}{m}\sum_{i=1}^{m}\left|\left|x^{(i)}\right|\right|^2} \tag{6}$$

The calculation for each number of selected principal components produced the following results, described in Table 3:

Table 3. Retained variance in dimension reduction process

Number of principal components	1	2	3
Retained variance	70.74%	90.43%	100%

As can be seen in Table 3, for two principal components new data vectors can be produced that contain about 90% of the variance of the original data, where the correlation between the two variables is negligible (due to the dimensionality reduction process). Therefore, we produce two new data features; S_1 - projection of the vector x on U1 and S_2 - projection of the vector x on the U2.

Following the dimension reduction, we now perform a regression with the single dependent variable being the coordination ability of the ith player, described by the ICA index, while the predictors are the newly found S1 and S2 features. In order to find the model coefficients, we used multiple linear regressions:

$$ICA(i) = 0.52118 + 0.090788 * S_1 + 0.054041 * S_2 \tag{7}$$

$$R^2 = 0.8733; F = 310.0768; p < 0.001; VAR_{error} = 0.0029 \qquad (8)$$

Where:

$$S1 = 0.610 * SR_{Closeness} + 0.518 * SR_{Equality} + 0.599 * SR_{Accession} \qquad (9)$$

$$S2 = -0.312 * SR_{Closeness} + 0.853 * SR_{Equality} + -0.418 * SR_{Accession} \qquad (10)$$

We can see from Eq. 8 the strong correlation between the individual coordination abilities and the strategic profile, while our correlation is based on only two variables, S_1 and S_2. This allows us to draw a surface that represents the model in a three-dimensional plane against the corresponding ICA values, as can be seen in Fig. 4.

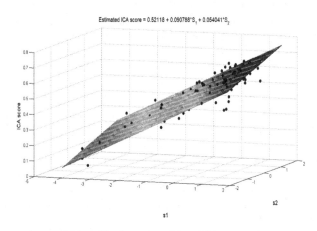

Fig. 4. The regression model in a 3D plane. A multi-variable regression model that describes the relationship between the three SR variables and the coordination ability represented by the ICA value of each player. The new model is presented as a surface in a 3D plane.

4   Conclusion

This research contributes to key issues pertaining to the process of decision-making in tacit coordination. First, to the best of our knowledge, this is the first study that demonstrates variability in players tacit coordination abilities. Second, to find the predominant selection rules that were applied during task performance, we devised the SR index that was computed for each individual player. This analysis hinted to the existence of two predominant selection rules: closeness and equality. This indicates that certain selection rules are preferred by players and are more useful for coordination than others in the context of a specific task. Third, we devised a novel method for constructing a strategic profile for each individual player. The method relies on the projection of the individual SR indices onto a 3D strategy space.

Finally, after demonstrating that there are differences between the coordination capabilities of the various players, and that a defined set of strategic profiles that characterize the players' choices could be extracted, we wanted to examine whether the strategic profile could serve as a good predictor of the player's coordination ability. For that purpose, we used multivariate correlation analysis and showed that there is a strong correlation between the player's strategic profile and its coordination ability. Previous research dealing with focal points examined a wide range of selection rules in their different games. However, in those studies no connection was demonstrated between individual coordination ability and the set of strategies implemented during the game as was done here.

After showing for the first time that people differ in their ability to succeed in tacit coordination games, many possible directions remain to be taken. First, it will be interesting to see if there are some objective traits that correlate with coordination ability. It might be one of the personality traits such as introversion or extroversion [11], a social value orientation (e.g. [12]), or any other trait that might provide information on the strategic profile at hand. Another interesting extension of the current research would be to consider tacit coordination games of divergent interests. That is, games in which the parties are required to coordinate their answers, however, different outcomes of the game may yield different utility to each of the players.

References

1. Bacharach, M.: Beyond Individual Choice: Teams and Frames in Game Theory. Wiley, New York (2006)
2. Binmore, K.: Playing for Real: A Text on Game Theory. Oxford University Press, Oxford (2007)
3. Schelling, T.C.: The Strategy of Conflict. Cambridge (1960)
4. Larrouy, L.: Revisiting methodological individualism in game theory: the contributions of schelling and bacharach (2015)
5. Nash, J.: Equilibrium points in n-person games. Proc. Natl. Acad. Sci. U. S. A. **36**, 48–49 (1950)
6. Mehta, J., Starmer, C., Sugden, R.: The nature of salience: an experimental investigation of pure coordination games. Am. Econ. Rev. **84**, 658–673 (1994)
7. Mehta, J., Starmer, C., Sugden, R.: Focal points in pure coordination games: an experimental investigation. Theory Decis. **36**, 163–185 (1994)
8. Hartigan, J.A., Hartigan, P.M.: The dip test of unimodality. Ann. Stat. **13**, 70–84 (1985)
9. Rousseeuw, P.J.: Silhouettes: a graphical aid to the interpretation and validation of cluster analysis. J. Comput. Appl. Math. **20**, 53–65 (1987)
10. Tukey, J.W.: Comparing individual means in the analysis of variance. Biometrics **5**, 99–114 (1949)
11. Eysenck, H.: The Biological Basis of Personality. Routledge, New York (2017)
12. Mizrahi, D., Laufer, I., Zuckerman, I., Zhang, T.: The effect of culture and social orientation on Player's performances in tacit coordination games. In: Proceedings of the Brain Informatics - International Conference, BI 2018, Arlington, TX, USA, 7–9 December 2018, pp. 437–447 (2018)

EEG-Based Driver Drowsiness Detection Using the Dynamic Time Dependency Method

Haolan Zhang[1,2(✉)], Qixin Zhao[1], Sanghyuk Lee[3],
and Margaret G. Dowens[4]

[1] The Center for SCDM, NIT, Zhejiang University, Ningbo, China
haolan.zhang@nit.zju.edu.cn
[2] Ningbo Research Institute, Zhejiang University, Hangzhou, China
[3] Xi'an Jiaotong-Liverpool University, Suzhou, China
sanghyuk.lee@xjtlu.edu.cn
[4] The University of Nottingham, Ningbo, China
Margaret.Dowens@nottingham.edu.cn

Abstract. The increasing number of traffic accidents caused by drowsy driving has drawn much attention for detecting driver's status and alarming drowsy driving. Existing research indicates that the changes in the physiological characteristics can reflect fatigue status, particularly brain activities. Nowadays, the research on brain science has made significant progress, such as the analysis of EEG signal to provide technical supports for real world applications. In this paper, we analyze drivers' EEG data sets based on the self-adjusting Dynamic Time Dependency (DTD) method for detecting drowsy driving. The proposed model, i.e. SEGAPA, incorporates the time window moving method and cluster probability distribution for detecting drivers' status. The preliminary experimental results indicates the efficiency of the proposed method.

Keywords: EEG pattern recognition · Drowsy driving detection · Dynamic time dependency · Brain informatics

1 Introduction

The increasing risk of traffic accidents caused by drowsy driving has become a crucial issue of road safety. The U.S.A national sleep foundation study shows indicated in the report that: "100,000 reported crashes are directly caused by driver fatigue each year in USA, which resulted in an approximate 1,550 deaths, 71,000 injuries, and $12.5 billion financial losses. In 2015, 2.3% of total crashes in USA were involving drowsy driving". Notably, drowsy driving has been a wide-spread phenomena in the world [1].

Based on our previous study on electroencephalogram (EEG) data analysis, we applied the newly proposed EEG recognition model, i.e. SEGPA, to drowsy driving detection applications. We conducted a concrete review on the formation mechanism, analysis method and fatigue state of EEG. The EEG spectrum of 16 channels was obtained and analyzed using conventional data analytical methods. In this paper, we study that the θ wave of the EEG signal, which can efficiently detect and recognize drivers fatigue state.

© Springer Nature Switzerland AG 2019
P. Liang et al. (Eds.): BI 2019, LNAI 11976, pp. 39–47, 2019.
https://doi.org/10.1007/978-3-030-37078-7_5

Currently, the major methods to identify fatigue driving include the subjective evaluation method, driver behaviour recognition and physiological signals analysis. The subjective evaluation method mainly recognizes a driver's external fatigue, aiming at the characteristics of personal feeling, response state and facial expression. The driver behaviour recognition method is to judge the driver's fatigue state by the control behavior during the driving process, such as the driver's blink frequency, vehicle trajectory and driving speed [2]. The physiological signals analysis method determines drowsy driving status by collecting physiological signals such as respiratory parameters, EEG and ECG [2]. Among them, fatigue driving recognition by collecting electroencephalogram (EEG) is recognized as the most accurate and objective method at present.

2 Related Work

Event-related Potential (ERP) in EEG signals is a real-time recording of EEG waves caused by stimulus events. The research methods of event-related potential (ERP) mainly include signal preprocessing, feature extraction and classification.

EEG signal is a very complex and regular weak signal. External noise interference will affect the accuracy of signal measurement. Therefore, only by eliminating external interference can we conduct better EEG signal analysis and research. Signal preprocessing is also an important step in EEG research. Existing research on EEG data processing and recognition has been conducted extensively in many areas, which include the following areas.

(1) EEG Signal preprocessing: Due to the large amount of data generated during EEG data collection, the first step is to reduce the amount of data. Usually, this step is based on experience or algorithm procedures to select electrons, and to remove the artifacts of EEG signal, which results in filtering EEG signals. Human EEG signal is a weak electrical signal, which is easily disturbed by various noises in the acquisition process. Common artifacts include blinking, ECG artifacts, motion interference and so on. These noises can be filtered out with a band pass filter. The selection of filter types depends on the characteristics of EEG signals. The frequency characteristic curve of Butterworth filter is the most flat. It is a monotone function of both pass band and stop band. The formula of amplitude square pair frequency for Butterworth filter is shown in Eq. 1. Chebyshev filter is a filter with some specific characteristics, which is uniformly distributed in the whole pass band, therefore the effect can be achieved with a lower order filter. The amplitude equation of Butterworth is listed as below.

$$|H(\omega)|^2 = \frac{1}{1 + \left(\frac{\omega}{\omega_c}\right)^{2n}} = \frac{1}{1 + \varepsilon^2 \left(\frac{\omega}{\omega_p}\right)^{2n}} \tag{1}$$

(2) EEG Feature extraction and classification methods: Current EEG feature extraction methods include power spectrum estimation, wavelet transform and independent component analysis. Basically, these methods can be divided into four categories: time domain analysis, frequency domain analysis, time-frequency analysis and

space-time joint analysis. Power spectrum estimation is a typical fast frequency domain analysis method. The power spectral density of a stationary random signal is estimated by using the given N sample data. Wavelet transform is a time-frequency signal processing method. Variable time-frequency windows were used to analyze the different frequency components of EEG signals. Independent Component Analysis (ICA) is a technique for separating source signals from multiple mixed signals based on statistical uncorrelation between sources.

The current major EEG classification methods are listed as follows. The Linear Discriminant Analysis (LDA) method establishes a probability density model for each category. The probability of each category in data can be calculated. The category corresponding to the maximum probability value is the category of input weight. The Artificial Neural Network (ANN) method has been applied in the EEG classification process that allows an accurate and efficient EEG recognition.

Nowadays, scientists have carried out specific research on fatigue identification methods. Doppler radar is used for the first time in the United States. Through complex signal processing, the "dozing driver detection system" has been developed for fatigue driving detection. The driver's fatigue state was judged by acquiring the data of driver's mental and emotional activities, blinking frequency and time [3].

Some researchers proposed a system to determine driver's fatigue status through analyzing eyes blinking speed and movement status [4]. The results show that the normal blinking frequency time is between 0.2 and 0.3 s. If the driver's eye blinking time reaches more than 0.5 s, it will more likely to cause accidents.

3 Driver Status Detection Based on DTD Method

3.1 Self-adjust Time Dependency for EEG Analysis

The SEGPA model introduced in [5] utilizes the clustering algorithm to generate initial EEG data clusters and utilizes the time dependency analysis method for EEG status recognition. In this paper, we further modified the SEGPA model through generating the EEG pattern recognition (PR) tree by using the differentiation values of two sequential EEG points instead of their original amplified voltage values in [5].

The clustering method has been developed to process various static data sets. The hierarchical clustering, model-based clustering as well as partition clustering methods have been adopted for time series clustering in EEG analysis. There are three major categories of time series classification according to the direct use of raw data, indirect use of features extracted from raw data or indirect use of models constructed from raw data. The correlation analysis of time series data is based on [6]. The model introduced in this paper constructs a generalization of the δ_j's, which is sensitive to the hypothesis of j-dependence in k-dimension, and is defined as follows [6]:

$$\delta_j^{[k]} = \frac{C_k - (C_j/C_{j-1})^{k-j} C_j}{C_k} = 1 - \left(\frac{C_j}{C_{j-1}}\right)^{k-j} \frac{C_j}{C_k} \tag{2}$$

where δ_j denotes dependencies that are the result of averages over regions of a map. C_k measures the probability that two vectors are within ε of each other in all their Cartesian coordinates.

The newly introduced model, i.e. SEGPA, merges and modifies the dependency measurement method introduced in [6]. This method calculates a part of the aggregated EEG time series data set with another aggregated EEG data of the same size in different time windows overlapping within a small size. The k-means method is applied to the clustering process, and the Minkowski distance is used to calculate the expression as follows:

$$D(X, Y) = (\sum_{i=1}^{n} |x_i - y_i|^p)^{1/p} \tag{3}$$

The advantages of using the Minkowski distance is that the distance calculation can be efficiently converted to other forms such as Manhattan distance and the Euclidean distance when p is 1 or 2, which can be expressed as follows:

$$d_{manhatta}(c_1, c_2) = \sum_{i=1}^{n} |x_i - y_i| \qquad d_{Euclidean}(c_1, c_2) = \sqrt{\sum_{i=1}^{n} (x_i - y_i)^2} \tag{4}$$

SEGPA model is based on K-means clustering method to cluster each EEG data separately. Therefore, the Manhattan distance and the Euclidean distance are both used in the clustering process based on different scenarios. However, in some high dimensional or more complicate EEG data forms, Chebyshev distance could be used, which can be derived based on the Minkowski distance as below:

$$\lim_{p \to \infty} \left(\sum_{i=1}^{n} |x_i - y_i|^p \right)^{\frac{1}{p}} = \max_{i=1}^{n} |x_i - y_i|. \tag{5}$$

The original sample EEE data set is shown on the left of Table 1. In order to further generate an effective EEG pattern, this paper further modified the SEGPA model by using the differentiation values instead of using the original EEG data sets. These data sets will be clustered using K-means classification method and further applied to the modified FP growth tree method described in [6]. The classification results are shown on the right of Table 1.

The following algorithm illustrates the EEG pattern recognition tree process used in SEGPA, namely SEGPA PR tree. Where PD(C) denotes Poisson Distribution of data set C, N_j denotes the initial node number.

Table 1. Converting original EEG data sets to time-series differential data Sets

Time ID	Original Data		Series ID	EEG Data	Clusters
0.01	79.4287		1	0	Cluster1
0.02	281.1286		2	201.6999	Cluster3
0.03	109.7588		3	171.3698	Cluster3
0.04	-133.0819	⇒	4	242.8407	Cluster3
0.05	93.44259		5	226.5245	Cluster3
0.06	354.4012		6	260.9586	Cluster3
0.07	137.6864		7	216.7148	Cluster3
0.08	52.40192		8	85.28448	Cluster2
......	......		......	......	......

Algorithm 1: Optimized SEGPA PR-tree Construction

Input: EEG time series data set C, Time elapsing t.
Output: PR-tree TR, EEG data pattern P(TR).
1 Calculate distribution of C, PD(C)$\longrightarrow F_i$ list (F_i list is in ascending order).
2 'Null' $\longrightarrow TR$, j=0, $N[\ ]$=$null$
3 **For** i = 0 **to** number(F_i)
4   **For** j = 0 **to** number(N_j)
4   **If** ($F_i \neq N_j[0]$) **then**
6              $F_i \longrightarrow TR$'s root
7   **Else**
8   **For** k = 0 **to** length($N_j[k]$)
9   **If** $F_{i+k} = N_j[k]$ **then**
10                     $F_i \longrightarrow TR$'s F_i with t variation
11   **End if**
12   **End for**
13   **End if**
14   **End for**
15 **End for**
16 **Return** TR

3.2 Experimental Design and Analysis

3.2.1 Experimental Settings

A number of EEG experiments have been conducted for drowsy driving tests. The experimental environment is: Windows 8 64-bit OS, Intel N3540 CPU, 4G RAM. The EEG recording time interval is 0.01 s for CONTEC KT88 used in this research.

The participants of the experiments are college students aged between 20 and 25 are in good health (studies have shown that young drivers are the most dangerous drivers). According to a survey conducted by relevant agencies in the United States, young drivers are prone to traffic accidents, including fatigue driving, which accounts for 50% of traffic accidents, among which young drivers aged 21–25 are the peak age. The participants were asked to remain calm or tired before the experiment was carried out; the subjects' physical health did not suffer from any sleep-related diseases. The experimental environment is shown in Fig. 1.

Due to the large amount of data collected from EEG signals, the EEG electron selection and initial artifact removal of the collected EEG signals are carried out first, and then the 2–30 Hz band-pass filter is processed by Chebyshev filter. Finally, the experimental data are extracted and adjusted. The signal pretreatment process is carried out by EDFbrowser software. EDFbrowser is a free and open source multi-platform data processing software, which supports *eeg*, *ecg*, *bdf* formats. EDFbrowser can realize the conversion of EEG file format, data processing, such as removing artifacts and filtering functions.

Fig. 1. EEG signal acquisition experiments

The detailed steps adopted in the experiments involve two main steps, which include the selection of EEG electrons step and filtering step. (1) In the EEG electron selection step, 16-channel signal acquisition was used in the experiment. The removal of unnecessary electrons that is not related to drowsy driving EEG data can be helpful to improve the accuracy and operation speed. Effective and invalid lead data are shown in Figs. 2-left and -right. Eight electron signals [FP1, FP2, F4, C3, C4, F8, T3, T4] were selected in our experiments.

(2) In the EEG signal filtering step. The EEG signals of each lead are processed by Chebyshev filter through 2–30 Hz band-pass filter in the interval. Figure 3 shows the original EEG data sets on the left and the filtered EEG data on the right. In EDFbrowser software, the power spectrum of EEG signals of different channels can be obtained by fast Fourier transform (FFT) (in the following example, the time period of the curve is 0:00:24).

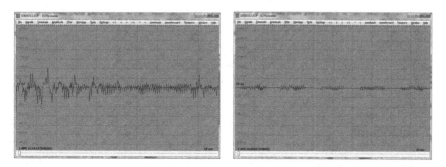

Fig. 2. F4-A2 electron signal (left) and F3-A1 electron signal (right)

Based on the experiments, the spectrum of EEG signal can be clearly obtained. Similarly, the EEG characteristic curves of the other six channels can be obtained in turn and the spectrum characteristics can be calculated. Finally, the piecewise power spectrum of EEG signal can be analyzed by statistical analysis of the spectrum characteristics of EEG signal.

Based on the processed EEG data sets, the optimized SEGPA model can generate a EEG pattern recognition tree [6] for drowsy driving pattern analysis. At the current stage, we applied the data curve fitting method and EEG power spectrum statistical method to the EEG drowsy driving detection process before using SEGPA pattern recognition (PR) tree. In the next stage, we will further improve the efficiency of the SEGPA PR tree. Table 2 shows the percentage of each EEG signal band in Drowsy Driver's EEG data [7].

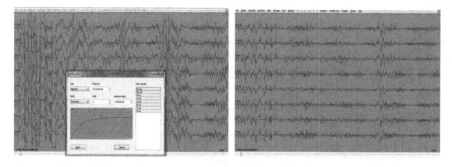

Fig. 3. The original EEG data (left) and filtered EEG data (right)

Table 2. Percentage of piecewise power spectrum

Electron	The percentage of each band in Drowsy Driver's EEG			
	δ	θ	α	β
FP1	15.36%	17.78%	19.32%	47.54%
FP2	15.21%	18.21%	19.38%	47.20%
F4	14.64%	20.81%	28.95%	35.60%
C3	10.31%	14.66%	28.81%	46.22%
C4	16.94%	25.01%	19.26%	38.79%
F8	12.77%	16.30%	25.49%	45.44%
T3	14.10%	19.56%	25.94%	40.40%
T4	13.14%	15.70%	20.32%	50.84%

4   Conclusion

This research addresses the specific problem of drowsy driving detection. We have summarized the research findings.

(1) Under the same condition, the proportion of EEG in different channels is different. For example, θ wave accounts for 14.66% in C3 channel and 25.01% in C4 channel.

(2) The proportion of power spectrum distribution of EEG in the same channel is different under different conditions or in different periods. Therefore, it can be assumed that the power spectrum distribution of human beings varies in different states.

(3) Research shows that theta wave in EEG signals can reflect fatigue state. This paper also confirms that fatigue state is related to theta wave change. The change of theta wave is most obvious in leads F4 and C4. It is verified that EEG signals can recognize fatigue state.

The future work will focus on improving the optimized SEGPA pattern recognition based on the current results.

Acknowledgement. This work is partially supported by Zhejiang Natural Science Fund (LY19F030010), Zhejiang Philosophy and Social Sciences Fund (20NDJC216YB), Ningbo Innovation Team (No. 2016C11024), National Natural Science Fund of China (No. 61572022). Ningbo Natural Science Fund (No. 83, chief investigator Haolan Zhang, Research on non-invasive BIC technology based on dynamic networks and machine learning methods, 2019).

References

1. National Cent. Stat. Analysis.: Drowsy driving 2015 - crash stats brief statistical summary. Report No. DOT HS 812 446, Washington, DC: National Highway Traffic Safety Administration (2017)

2. Huang, J., Zhang, L., Xu, J.: Research on EEG-based fatigue driving. Ergonomics **4**, 36–40 (2015)
3. Wang, W.: Research on Driver Fatigue Detection System. Thesis. South China University of Technology (2010)
4. Akrout, B., Mahdi, W.: A blinking measurement method for driver drowsiness detection. In: Burduk, R., Jackowski, K., Kurzynski, M., Wozniak, M., Zolnierek, A., (eds.) Proceedings of the 8th International Conference on Computer Recognition Systems CORES 2013. Advances in Intelligent Systems and Computing, vol. 226. Springer, Heidelberg (2013). https://doi.org/10.1007/978-3-319-00969-8_64
5. Zhang, H.L., Xue, Y., Zhang, B., Li, X., Lu, X.: EEG pattern recognition based on self-adjusting dynamic time dependency method. In: Proceedings of ICDS. Lecture Notes on Computer Science. Springer (2019)
6. Nie, D., Fu, Y., Zhou, J., Fang, Y., Xia, H.: Time series analysis based on enhanced NLCS. In: Proceedings of ICIS, pp. 292–295. IEEE Press (2010)
7. Zhao, Q.: Exploring Fatigue Driving Recognition and Early Warning Based on EEG Data Analysis. Thesis, Zhejiang University, NIT (2017)

Human Information Processing Systems

Route Adjustment of Functional Brain Network in Mental Arithmetic Using Task-Evoked FMRI

Xiaofei Zhang[1,2,4,5], Yang Yang[3,4,5,6], Ruohao Liu[1,4,5], and Ning Zhong[1,4,5,6(✉)]

[1] Faculty of Information Technology, Beijing University of Technology,
Beijing 100124, China
julychang@just.edu.cn
[2] School of Computer, Jiangsu University of Science and Technology,
Zhenjiang 212003, China
[3] Department of Psychology, Beijing Forestry University,
Beijing 100083, China
[4] Beijing International Collaboration Base on Brain Informatics
and Wisdom Services, Beijing 100124, China
[5] Beijing Key Laboratory of MRI and Brain Informatics, Beijing 100124, China
[6] Department of Life Science and Informatics,
Maebashi Institute of Technology, Maebashi, Gunma 371-0816, Japan

Abstract. A large number of studies on altered functional brain network tend to focus only on the alternation in topological metric of functional connectivity, rather than on the details of graph adjustment that cause topological metric changes, such as significant adjusted route and the nodes on it. In this paper, we first used the brain atlas of Dosenbach to generate the functional brain networks of the 21 participants recruited in the mental arithmetic experiment. Then, the nodal efficiency of each brain region in the network were calculated and statistically compared between mental arithmetic cognitive states. The brain regions with significant alternation in nodal efficiency were taken as seeds for searching adjusted routes. The brain regions that have significant changes in network efficiency with the seed nodes were considered as destined nodes of the relative seed nodes. Finally, the details of two adopted indicators on altered functional brain network by comparing the adjusted route between the two endpoints of the adjusted route were given and used as clues for the better understanding of the cognitive pattern of mental arithmetic. In this paper, the average number of adjusted routes contributed by brain region is used to indicate the degree of contribution of the brain region to the route adjustment, and the interaction degree within specific network is indicated by the density of adjusted routes. The results show that both indicators of fronto-parietal network is significantly higher than that of other networks, which indicates the brain regions and the routes within fronto-parietal network are the most active. In summary, the method proposed in this paper provides a new perspective to study the causes of functional brain network alternation in mental arithmetic. However, due to the participants' variation of adjusted routes and the nodes on it, a better understanding of these functional brain network alternation for individual participant with the proposed method needs more in-depth research.

© Springer Nature Switzerland AG 2019
P. Liang et al. (Eds.): BI 2019, LNAI 11976, pp. 51–61, 2019.
https://doi.org/10.1007/978-3-030-37078-7_6

Keywords: Mental arithmetic · Functional brain network · Nodal efficiency · Route adjustment · Fronto-Parietal network

1 Introduction

Mental arithmetic cognitive experiment can be used to study the brain's underlying mechanisms when the human brain performs arithmetic calculations. From the point of view of individual brain region, anterior central gyrus and lobule of posterior parietal lobe were argued to be tightly related to mental arithmetic, and the centers of processing and working memory [1]. Brain regions of bilateral intraparietal sulci and inferior occipital gyrus were also found more activated compared with the processing of arithmetic principles [2]. Some studies have even studied the mental arithmetic by considering more subdivided mental arithmetic tasks, trying to understand more aspects about the cognitive mechanism of mental arithmetic. Left inferior frontal gyrus, middle portion of dorsolateral prefrontal cortex, and supplementary motor area were found to be more active where the recruited participant performed mental subtraction task than mental addition task [3].

Functional brain network may be mathematically modeled as graphs and be used to reveal the properties of human's cognitive functions, including mental arithmetic. Under different mental arithmetic cognitive tasks, topological metrics of the connectomes show different trends in network efficiency [4]. The arithmetic ability of adults were found to depend on the fronto-parietal control network (FPCN), in which the parietal regions including superior parietal lobule (SPL) and inferior parietal lobule (IPL), and frontal regions including inferior frontal gyrus (IFG), middle frontal gyrus (MFG) and left superior frontal gyrus [5–8]. Therefore, studying mental arithmetic from the perspective of functional brain network is a potential way to understand its underlying mechanism.

Recently, an increasing number of researches have begun to focus on dynamic changes of functional brain network instead of limiting to the static functional brain networks. This new research perspective usually needs to consider the size of the time window used to extract sequential (or rolling) fMRI BOLD signals carefully. For example, time-resolved network analysis of fMRI data has been used to demonstrate about the traverses of human brain between functional states that maximize either segregation into tight-knit communities or integration across otherwise disparate neural regions [9]. The dynamic configuration of functional brain networks of three distinct cognitive visual tasks in recognition memory and strategic attention was studied, by dividing regional BOLD time series into variable sized time window [10]. The dynamic reconfiguration of human brain networks during learning was examined by identifying dynamic changes of modular organization spanning multiple temporal scales [11]. Many technologies related to dynamic functional brain network have been used to study the dynamics of 334 participants' fMRI data in working memory task, that is, n-back task [12]. The dynamic reconfiguration of supplementary motor area network is also supposed to be related with the imagined music performance, and the network show an increased connectivity compared to that of resting state [13]. The dynamics of such network reconfiguration will eventually manifest themselves in changes in topological metrics, and the root cause lies in the changes in connections and routes in the functional brain network.

In this paper, we used graph theoretic analysis to calculate the nodal efficiencies of brain regions according to brain atlas of Dosenbach, and searched the significantly adjusted routes during mental arithmetic experiment. Our initial hypothesis is that during the mental arithmetic experiment, there are significant adjustments on the critical routes in the brain network under different cognitive states, and the adjustments of these routes are the potential clues to discover the mental arithmetic mechanism. To verify this hypothesis, correlation matrix of each cognitive state (addition and subtraction) in mental arithmetic experiment was first constructed. Next, the nodal efficiency of each brain region was calculated, and the brain regions that have significant differences between the two cognitive states were found with statistical tests. Finally, these brain regions with significant differences were served as the source node for finding the adjusted routes and the corresponding destined node, by statistically testing the route length and efficiency to other brain regions. Figure 1 shows the main steps of the research method with an overall perspective in this paper. The detailed contents will be explained in the following sections.

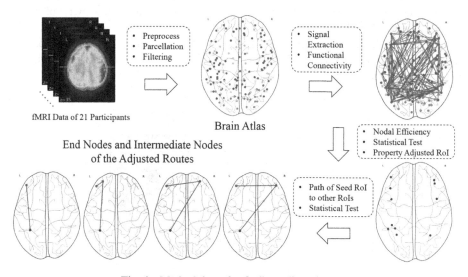

Fig. 1. Methodology for finding adjusted route

2 Materials and Methods

2.1 fMRI Data Acquisition and Preprocessing

The mental arithmetic experiment for collecting fMRI data in this paper is a psychological experiment involving multiple mental arithmetic cognitive tasks, including addition, subtraction, number matching, and resting state. The experiment is block designed and the emergence of different experimental conditions is pseudo-random arranged [4]. The main focus here is on the route adjustment of the functional brain network between cognitive states of addition and subtraction. 21 Chinese university graduate students without significant statistical difference were recruited in this experiment and the fMRI data were collected by using a 3.0 T MAGNETOM Trio Tim from Siemens Medical Systems in Erlanger, Germany.

The collected fMRI data were preprocessed with the steps including slice timing correction, realignment, spatial normalization and smoothing. Considering that some literatures argue that the strategy of preprocessing may impact the following analysis result, the methods and parameters used here were well thought out considering the impact of the choice of preprocessing strategy on the analysis results of fMRI data.

2.2   Construction of Functional Brain Networks

The functional brain networks are constructed using the brain atlas of Dosenbach [14], which consists of six subnets with a total of 160 brain regions. We averaged all the voxel BOLD signals in the 5 mm radius around the MNI coordinates of each brain region as the timing signals. Next, for each cognitive state of each participant, we calculated the Pearson correlation coefficient for each pair of brain regions and obtained a 160*160 correlation coefficient matrix for each cognitive state of each participant. Finally, the binary matrix of maximized intrinsic functional connectivity of each cognitive state of each participant was calculated with a threshold at P < 0.05 (FDR corrected), and these binary matrices were used as the functional brain network for the following topological property calculation. Figure 2 shows the main steps of functional brain network construction.

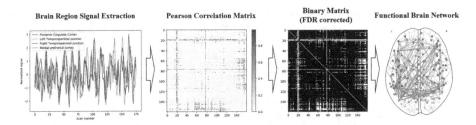

Fig. 2. The workflow of functional brain network construction

2.3   Topological Property Calculation

We use network efficiency [15] to evaluate the topology properties of functional brain network of mental arithmetic in different cognitive states. Considering the functional brain network as a network with communication and interaction between brain regions, the efficiency of the network is a measure of how efficiently it exchanges information. The efficiency between two brain regions is defined as Eq. (1).

$$Efficiency(i, j) = \frac{1}{D(i, j)} \tag{1}$$

From the perspective of graph theory, the efficiency between node i and node j is the reciprocal of the distance between them. Nodal efficiency [16] is an effective nodal measures of brain network efficiency concerning specific nodes of the network, and is defined as Eq. (2)

$$NodalEfficiency(i) = \frac{1}{n-1} \sum\nolimits_{i \neq j \in G} Efficiency(i, j) \qquad (2)$$

where the nodal efficiency of node i is the average of its efficiency with all other nodes in the graph.

By statistically testing the nodal efficiency of each brain region between the cognitive states of addition and subtraction, it is found that there are 25 brain regions in the brain atlas of Dosenbach with significant differences between the two cognitive states. Figure 3 shows the nodal efficiencies of the 21 brain regions in the fronto-parietal network of Dosenbach under the two cognitive states. These brain regions and their nodal efficiencies are given in Table 1, and among the 25 brain regions where significant differences exist, 11 from fronto-parietal, 6 from cingulo-opercular, 3 from sensorimotor, 2 from occipital, and 3 from cerebellum.

Table 1. RoIs with significant difference between addition and subtraction

RoI ID	RoI name	Network name	Nodal efficiency		P-value
			Addition	Subtraction	
38	vent aPFC	fronto-parietal	0.824648098	0.783418189	0.003326649
39	vlPFC	fronto-parietal	0.830388340	0.784666068	0.001689692
40	dlPFC	fronto-parietal	0.831536388	0.808824998	0.030589268
44	dlPFC	fronto-parietal	0.789707497	0.747928522	0.001418427
46	dFC	fronto-parietal	0.843266447	0.818308875	0.006107537
47	dFC	fronto-parietal	0.834730957	0.787760807	0.000679832
48	IPL	fronto-parietal	0.817510233	0.786163522	0.015452623
50	post parietal	fronto-parietal	0.835130279	0.807527204	0.034776711
51	IPL	fronto-parietal	0.821403614	0.779325147	0.003207047
54	IPS	fronto-parietal	0.827643007	0.798043326	0.019812870
55	IPS	fronto-parietal	0.839273235	0.799440950	0.003046980
60	ant insula	cingulo-opercular	0.820854547	0.793201557	0.032486453
62	ant insula	cingulo-opercular	0.816711590	0.787760807	0.025307219
64	mFC	cingulo-opercular	0.859389039	0.834980533	0.025509594
65	vFC	cingulo-opercular	0.821104123	0.794549266	0.043066071
83	parietal	cingulo-opercular	0.840970350	0.806129580	0.044825528
85	angular gyrus	cingulo-opercular	0.808275931	0.766846361	0.004153707
96	mid insula	sensorimotor	0.780772686	0.747728861	0.024537265
110	precentral gyrus	sensorimotor	0.809673555	0.772936009	0.011233172
120	sup parietal	sensorimotor	0.777078966	0.740940401	0.013575244
121	occipital	occipital	0.803683738	0.766746531	0.022322227
125	occipital	occipital	0.836777478	0.811570330	0.048739804
146	lat cerebellum	cerebellum	0.834830788	0.803783568	0.029087620
149	inf cerebellum	cerebellum	0.850354398	0.820405311	0.008587068
159	inf cerebellum	cerebellum	0.843166617	0.818109214	0.024326539

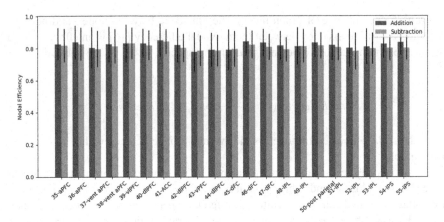

Fig. 3. Nodal efficiency of the RoIs in Dosenbach's fronto-parietal network

2.4   Searching Significant Adjusted Routes

After getting the brain regions whose nodal efficiencies are significantly changed under different mental arithmetic cognitive states, we used these brain regions as the original node and search the destined nodes where the route between them are significantly adjusted. The main steps of the process are shown in Fig. 4.

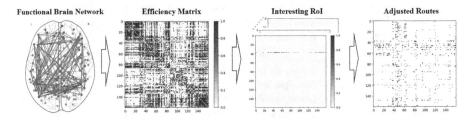

Fig. 4. The workflow of searching adjusted route

The first step is to calculate the network efficiency between each pair of brain regions of the functional brain network under a specific mental arithmetic cognitive state for each participant, and these network efficiencies are arranged as an efficiency matrix. The second step is to use a brain region with significant changes in nodal efficiency as seed node or origin node. Take the 38th brain region 'vent aPFC' of Dosenbach as an example, the 38th row of the efficiency matrix represent the network efficiencies between 'vent aPFC' and other brain regions. Hence, two cognitive states with 21 participants result 42 rows. That is, the two mental arithmetic cognitive states for this brain region have 21 row vectors respectively. A statistical test is performed between the cognitive states of addition and subtraction for each element in the rows to see if there is a significant difference in the network efficiency. The second step is repeated for each seed brain region and generate a binary matrix, named 'adjusted routes', at last. The cell with true value in the matrix indicates a significant change in network efficiency between the two nodes, while the false value indicates no significant change in network efficiency.

2.5   The Nodes on the Adjusted Route

After obtaining the matrix of 'adjusted routes', the destined brain regions with a significant adjusted in route from the source brain region can be seen. Similarly, in the case of brain region 'vent aPFC', the column corresponding to the black cell in the 38th row of the matrix is the destined brain region corresponding to the significant adjusted route between cognitive states of addition and subtraction.

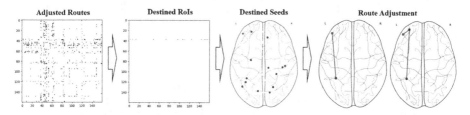

Fig. 5.   The workflow of searching the nodes on the adjusted route (Color figure online)

As shown in Fig. 5, the red node is the source brain region 'vent aPFC', and the blue nodes are the destined seeds with a significant change in the route from source brain region. If one of the destined seeds is selected as the target node, the intermediate nodes between the source node and the target node can be calculated and plotted in green. It should be noted that the search of the seed nodes is based on statistics. Therefore, the actual route between the same pair of source node and the target node is always different in the functional brain network of different participants, which provide the possibility to study the difference of individual participant.

3   Results

In the matrix of 'adjusted route', the significantly adjusted routes between the two mental arithmetic cognitive states are given. Since the matrix of 'adjusted route' is a symmetric, all the column vectors or row vectors of the matrix are summed to obtain the number of adjusted routes how much each brain region participates in. If these brain regions are grouped according to the functional networks defined in Dosenbach, then the number of involved adjusted route of each brain region is summed, and the number of involved adjusted route of each functional network in Dosenbach is obtained. The result is shown in Fig. 6(a), the average number of adjusted routes participating in each brain region in the fronto-parietal network is 14.534. In the other five networks, the average number of adjusted routes involved in each brain region was 9.176, 9.5, 8.152, 7.7272, and 8.111, respectively. From the results, it is obvious that the brain regions in fronto-parietal network involved with adjusted routes account for the most in the process of mental arithmetic cognition. This means the brain regions in fronto-parietal network are the most active compared to the ones in other network, and this discovery is consistent with the consensus that the arithmetic ability of adult is dependent on fronto-parietal network.

We observed the route adjustment between the brain regions within the same network, where the two endpoints of the adjusted route belong to the same network. Then, by dividing the number of adjustment routes within same network by all possible edges in the network, the density of route adjustment within individual network can be obtained. According to the results shown in Fig. 6(b), it can be seen that the density of route adjustment in the fronto-parietal network is the most active, reaching 11.43%. The densities of route adjustment within the networks of cingulo-opercular, sensori-motor, and occipital are less active, but still reaching 8.87%., 9.47%, 8.66% respectively. The densities of route adjustment in the network of default and cerebellum are the least active, 4.28% and 1.31%, respectively.

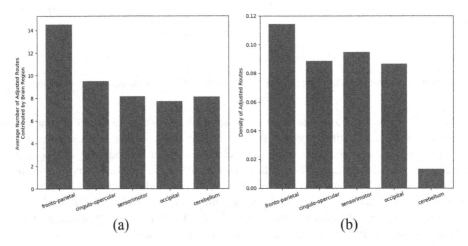

(a)   (b)

Fig. 6. Average number of adjusted routes contributed by brain region and density of adjusted routes in each network

4 Discussions

By applying graph theoretical analysis and statistical analysis on the task-evoked fMRI data collected in the mental arithmetic experiment, we first studied the significantly adjusted nodal efficiencies of the brain regions between the two cognitive states of addition and subtraction. It was found that there were significant differences in brain regions mainly distributed in the fronto-parietal network, and this result is consistent with the view that adult's mental arithmetic relies mainly on the fronto-parietal net-work. Then, taking the brain regions which have significant differences in nodal effi-ciency as seeds, the network efficiency with significant changes was calculated, and the endpoints of the adjusted routes were obtained. Calculating the mean of the adjusted routes involved in these endpoints, it is found that this indicator belonging to the endpoint of fronto-parietal is also significantly higher than other networks, indicating that the brain regions of fronto-parietal have higher contributions in activities participating in route adjustment. Finally, we calculated the ratios of adjusted routes in the two cognitive states within each network of Dosenbach, and this indicator of fronto-parietal is still the highest. In addition, the indicator of cingulo-opercular, sensorimotor, and occipital networks are also more active than that of default and cerebellum, which is consistent with the experimental paradigm itself, because the participants need cognitive control to perform memory extraction, visually read mental arithmetic problems, and answer questions by pressing a button.

In this paper, the brain regions, contributing to the significantly adjusted routes of functional brain network in mental arithmetic, and their distributions are studied. The found relationship between fronto-parietal network and mental arithmetic is also consistent with the conclusions of mainstream research. The universal of the approach will be applied and verified with more task-evoked cognitive experimental fMRI data, and we try to find more essential dynamics of functional brain network in diverse brain cognitive activities with a systematic strategy including (1) collecting and analyzing more fMRI data through systematic cognitive experiment design based on the brain informatics methodology [17], (2) using the conceptual model of the Data-Brain [18] to represent the concepts related to mental arithmetic cognition, and then to study the cognitive mechanism from the perspective of knowledge graph, (3) using the W2T [19] framework to conduct a broader range of mental arithmetic cognitive research and service development in an open form as WaaS [20].

5 Conclusions

Some overall and local indicators of functional brain network are adjusted during the mental arithmetic process of addition and subtraction. The adjusted indicators are generated by more detailed topological changes. Many studies on functional brain network tend to focus only on the indicator itself, rather than the cause of the change in the indicator. Nodal efficiency is one of them, and although it has portrayed changes at nodal level, many studies have not explored the reasons for this change. Based on this hypothesis, we firstly found the brain regions where the nodal efficiency is significantly different between the two cognitive states of mental arithmetic. Then we searched the significantly adjusted routes with network efficiency, which is the reciprocal of the path length, and then explored the routes and corresponding endpoints that cause the nodal efficiency changed significantly. The result is consistent with the idea that adult's ability of mental arithmetic relies on fronto-parietal network. In addition, the paper also offers a proposed method about how to further analyze the causes of route adjustment and the brain regions through which the route passes, but it is more in-depth because of the variations of the route adjustment in different participants. The analysis and effective results need more works to describe the functional brain network dynamics of the mental arithmetic process from a more comprehensive perspective.

Acknowledgements. This work was supported by grants from the National Natural Science Foundation of China (61420106005), the Science and Technology Project of Beijing Municipal Commission of Education (KM201710005026), and the JSPS Grants-in-Aid for Scientific Research of Japan (19K12123).

References

1. Wang, M., Wang, L.: Localization of the brain calculation function area with MRI. Chin. Sci. Bull. **46**(22), 1889–1892 (2001)
2. Liu, J., Zhang, H., Chen, C., et al.: The neural circuits for arithmetic principles. Neuroimage **147**, 432–446 (2016). (Complete)
3. Yang, Y., Zhong, N., Friston, K., et al.: The functional architectures of addition and subtraction: network discovery using fMRI and DCM. Hum. Brain Mapp. **38**(6), 3210–3325 (2017)
4. Zhang, X., Yang, Y., Zhang, M.-H., Zhong, N.: Network analysis of brain functional connectivity in mental arithmetic using task-evoked fMRI. In: Wang, S., et al. (eds.) BI 2018. LNCS (LNAI), vol. 11309, pp. 141–152. Springer, Cham (2018). https://doi.org/10.1007/978-3-030-05587-5_14
5. Arsalidou, M., Taylor, M.J.: Is 2 + 2 = 4? meta-analyses of brain areas needed for numbers and calculations. Neuroimage **54**(3), 2382–2393 (2011)
6. Dehaene, S., Cohen, L.: Cerebral pathways for calculation: double dissociation between rote verbal and quantitative knowledge of arithmetic. Cortex **33**(2), 219–250 (1997)
7. Klein, E., Moeller, K., Glauche, V., et al.: Processing pathways in mental arithmetic—evidence from probabilistic Fiber tracking. PLoS ONE **8**(1), 1–14 (2013)
8. Klein, E., Suchan, J., Moeller, K., et al.: Considering structural connectivity in the triple code model of numerical cognition: differential connectivity for magnitude processing and arithmetic facts. Brain Struct. Funct. **221**(2), 979–995 (2016)
9. Shine, J.M., Bissett, P.G., Bell, P.T., et al.: The dynamics of functional brain networks: integrated network states during cognitive function. Neuron **92**(2), 544–554 (2015)
10. Telesford, Q.K., Lynall, M.E., Vettel, J., et al.: Detection of functional brain network reconfiguration during task-driven cognitive states. Neuroimage **142**, 198–210 (2016)
11. Bassett, D.S., Wymbs, N.F., Porter, M.A., et al.: Dynamic reconfiguration of human brain networks during learning. Proc. Nat. Acad. Sci. U. S. Am. **108**(18), 7641–7646 (2011)
12. Braun, U., Schäfer, A., Walter, H., et al.: Dynamic reconfiguration of frontal brain networks during executive cognition in humans. Proc. Nat. Acad. Sci. U.S. Am. **112**(37), 11678–11683 (2015)
13. Tanaka, S., Kirino, E.: Dynamic reconfiguration of the supplementary motor area network during imagined music performance. Front. Hum. Neurosci. **11**, 1–11 (2017)
14. Dosenbach, N.U.F., Nardos, B., Cohen, A.L., et al.: Prediction of individual brain maturity using fMRI. Science **329**(5997), 1358–1361 (2010)
15. Latora, V., Marchiori, M.: Efficient behavior of small-world networks. Phys. Rev. Lett. **87**(19), 198701–198704 (2001)
16. Hilger, K., Ekman, M., Fiebach, C.J., et al.: Efficient hubs in the intelligent brain: nodal efficiency of hub regions in the salience network is associated with general intelligence. Intelligence **60**, 10–25 (2016)
17. Zhong, N., Yau, S.S., Ma, J., et al.: Brain informatics-based big data and the wisdom web of things. IEEE Intell. Syst. **30**(5), 2–7 (2015)

18. Zhong, N., Chen, J.: Constructing a new-style conceptual model of brain data for systematic brain informatics. IEEE Trans. Knowl. Data Eng. **24**(12), 2127–2142 (2012)
19. Zhong, N., Ma, J.H., Huang, R.H., et al.: Research challenges and perspectives on wisdom web of things (W2T). J. Supercomput. **64**(3), 862–882 (2013)
20. Chen, J., Ma, J., Zhong, N., et al.: Waas: wisdom as a service. IEEE Intell. Syst. **29**(6), 40–47 (2014)

Study on the Connectivity of Language Network in Word Reading and Object Recognition Based on tfMRI

Xiang He[1], Xiaofei Zhang[2,3,5,6], Yang Yang[4,5,6,8], Ting Wu[7], and Ning Zhong[2,5,6,8(✉)]

[1] School of Foreign Language, Jiangsu University of Science and Technology, Zhenjiang 212003, China
hexiang@just.edu.cn
[2] Faculty of Information Technology, Beijing University of Technology, Beijing 100124, China
[3] School of Computer, Jiangsu University of Science and Technology, Zhenjiang 212003, China
[4] Department of Psychology, Beijing Forest University, Beijing 100083, China
[5] Beijing International Collaboration Base on Brain Informatics and Wisdom Services, Beijing 100124, China
[6] Beijing Key Laboratory of MRI and Brain Informatics, Beijing 100124, China
[7] Mental Health Education Center, Jiangsu University, Zhenjiang 212003, China
[8] Department of Life Science and Informatics, Maebashi Institute of Technology, Maebashi, Gunma 371-0816, Japan
zhong@maebashi.ac.jp

Abstract. Brain structural connectivity is the foundation of its functionality. To understand the brain abilities, studying the relation between structural and functional connectivity is essential. This study aims to investigate the structure connectivity and the information exchange efficiency of the brain language network under different task modality. Using a public database from the open project "Word and Object Processing" shared from the OpenfMRI website (https://openfmri.org/dataset/), this study analyzed task-state fMRI data of 45 subjects with high temporal and spatial resolution. Based on the topological characteristics of language network connection efficiency, this study investigated the structure connectivity and the information exchange efficiency of the brain language function network under different tasks. The result findings show that the structure connectivity and the information exchange efficiency of the brain language network are strongly affected by the task modality. The task of written words reading, compared with other tasks, gets more language network nodes involved and thus the strongest structural connectivity will be activated and the highest information exchange efficiency will be achieved.

Keywords: Language network · Stimulus modality · Structure connectivity · Information exchange efficiency

© Springer Nature Switzerland AG 2019
P. Liang et al. (Eds.): BI 2019, LNAI 11976, pp. 62–71, 2019.
https://doi.org/10.1007/978-3-030-37078-7_7

1 Introduction

Functional networks are fundamental to the brain cognitive processes in humans. Conventional functional neuroimaging studies mainly focused on the localization of the functional region-of-interest (fROI) activated by experimental manipulation, which were designed at localizing the specific neuroanatomical region functionally based on its response properties. In most cases, this involves collecting scans in which participants perform a different task solely for the purpose of functionally identifying the anatomical region. To localize the specific neuroanatomical region functionally based on its response properties, many functional neuroimaging studies have investigated the sensitive voxels by using functional data to identify a region-of-interest and define it based on the cluster of voxels within a given anatomical area activated by a particular contrast [1–6].

Human brain is a complex dynamic system, consisting of various subsystems of brain network, and each region may subserve a number of various working processes. Temporal correlations in fMRI data (functional connectivity) are closely related to anatomical connectivity patterns [6], for instance, an anterior and inferior prefrontal region may be involved selectively in semantic processing. Each mental task involves a series of cognitive processes, including retrieving, maintaining, monitoring and manipulating semantic or concept representations stored elsewhere in the brain network. Thus, the successful completion of cognitive identification task will involve the coordination and integration of various parts of the brain neural network as a whole. Nevertheless, previous studies have mainly focused on investigating the response properties of specific neuroanatomical regions in isolation, and few studies have tried to report the consistency of the activation with intra-subjects, nor the possible total connectivity and interregional information-exchanging capability in the brain neural network for the cognitive processing tasks.

Brain networks are invariably complex, connected by anatomical tracts and various functional associations. By sharing common features of networks from other biological and physical systems, the brain connectivity datasets may be characterized using complex network methods. In this paper, we will use topological property of global efficiency to measure the collected fMRI data of the 45 healthy adult subjects participating in the experiments with four types of stimulus input (written English words and pictures of objects). Here, our goal is to explore how the integration strength and efficiency of human brain's language network depend on the stimulus types. Thus, our hypothesis is that subjects may use different processing strategies for the identification of pictures vs. words (e.g. visual identity vs. lexical or phonological identity), resulting in different brain responses, reflected as different collaborating strength and efficiency of human brain language network, for words and pictures during the one-back identity task.

In order to verify this hypothesis, we calculated the functional connectivity of the language network under the different experiment conditions with four categories of visual stimuli, namely written words, consonant letter string, pictures of common object, and scrambled pictures of the same objects, then calculated the topological property of global efficiencies of language network at 100 graph densities for each subject, finally the significance of difference between different experiment conditions were statistically tested. The overall research workflow is depicted in Fig. 1 and the detailed content are described in the following sections.

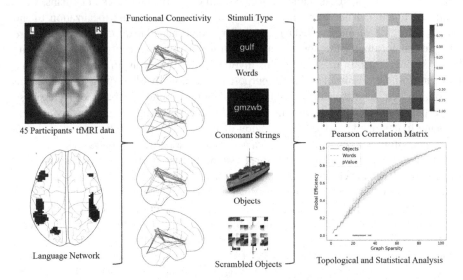

Fig. 1. Research workflow of this paper

2   Materials and Methods

2.1   Experiment Data

This study involves a re-analysis of fMRI data from Duncan et al. [7], so that we could focus on how the language neural network of human brain is influenced by stimulus modality. We obtained the experiment data from an opened project 'word and object processing' shared on the website of OpenfMRI (https://openfmri.org/dataset/). The dataset includes experimental process information, T1 anatomical MRI image, and functional MRI images of 49 subjects. During processing of the fMRI data, it was found that the fMRI data of two subjects could not be properly decompressed, and that of other two subjects cannot be preprocessed correctly with the software of SPM (Statistical Parametric Mapping [8]). Therefore, this study can only use the data of the rest 45 subjects to conduct analysis, and the original paper also mentioned the use of 45 subjects of the data for research.

2.2   Experiment Design

The experiment is designed to investigate whether the human brain's functional activations can be selectively elicited for words reading and object recognition in the language network with two runs of functional localizer scans. A one-back 'identity' task was used with four categories of visual stimuli: written words, pictures of common objects, scrambled pictures of the same objects, and consonant letter strings. The entire experimental process consisted of two runs, as shown in the Fig. 2(a), and the subjects were allowed to have a short break during the intervals. Subjects were instructed to press a button if the identical stimulus was repeated. To maximize the statistical sensitivity, a block design was used and each block consisted of 16 trials from a single category presented one every second, as shown in Fig. 2(b).

The time interval between two successive block groups is 16 s, which means subjects viewed a fixation cross for 16 s in between blocks. All the stimuli were cut equally into two block groups and each one was listed in the counter-balanced order across subjects. As shown in Fig. 2(c), each block group consists of four category of stimuli (written English words, pictures of object, scrambled object pictures, and consonant strings), and there were 192 stimuli per category including targets in total. As shown in Fig. 2(d), a block contains 16 trials, and the duration of one trial is be about 1 s. The change in time was due to the fact that the subject needs to respond to the stimulus material. A trial began with a 650 ms fixation cross, followed by the stimulus for 350 ms. In the one-back identify task, subjects were instructed to press a key pad if the stimulus was identical to the preceding one (e.g. dog, dog). The one-back identity task was used here to achieve the variety of stimulus category without changing the task to maintain a constant cognitive set.

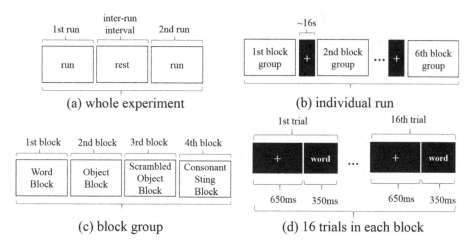

Fig. 2. Experiment paradigm

2.3 Data Preprocessing

There are two procedures in the workflow of the fMRI data preprocessing in this paper. The first stage is made up of four steps (slice timing, realignment, normalization, smooth) and conducted by using software of SPM, and the second stage is comprised of two steps (detrend and filtering), which was accomplished by using a Python API of NiLearn [9]. For the step of slice timing, we chose the 17th slice as the reference slice since the initial fMRI volume has 35 slices. The parameter of TR and TA are 3 and 2.914 s respectively. As for the steps of realignment, normalization and smooth, we used the default parameters of the SPM and generated 3 mm × 3 mm × 3 mm nifti format volume file. The two steps of the second phase were completed by calling one function from NiLearn package named nilearn.signal.clean. The low pass and high pass of filtering were set as 0.1 Hz and 0.01 Hz respectively.

2.4 Functional Connectivity Calculation

The calculation of the functional connectivity needs to use a brain atlas to define the functional brain network. Since this paper focus on the language network during word and object processing, we referred the definition of language network of Willard-499 [10]. There are 14 functional networks defined in Willard-499, and language network consists of 9 regions, as shown in Fig. 3. The region of Willard-499 is defined anatomically, hence, the number of voxels of each region is different. The RoI ID, the number of voxels, the center MNI coordinate, and the center voxel coordinate of each region in language network are given in Table 1.

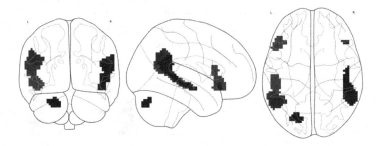

Fig. 3. Language network defined in Willard-499

Table 1. RoIs of language network in Willard-499 Atlas.

RoI ID	Number of voxels	MNI coordinate of RoI center (mm)	Voxel coordinate of RoI center (3*3*3)
396	155	(−46.86, 23.52, −4.37)	(45.62, 49.84, 22.54)
397	101	(−50.76, −32.73, −6.92)	(46.92, 31.09, 21.69)
398	228	(−53.49, −55.70, 15.12)	(47.83, 23.43, 29.04)
399	52	(−53.48, −50.94, 30.58)	(47.83, 25.02, 34.19)
400	118	(−48.53, −59.67, 26.16)	(46.18, 22.11, 32.72)
401	13	(49.15, 26.08, −12.69)	(13.62, 50.69, 19.77)
402	211	(58.52, −50.84, 17.03)	(10.49, 25.05, 29.68)
403	109	(53.42, −31.13, −6.63)	(12.19, 31.62, 21.79)
404	72	(−20.13, −81.58, −36.92)	(36.71, 14.81, 11.69)

From the spatial perspective, we extracted the fMRI signals of each subject according to the voxel coordinates contained in each brain region defined by the language network, and average the time series of all voxels contained in the same brain region as the time series for subsequent processing. From the temporal perspective, we divide the time series of each brain region according to the sequence information of experiment conditions. Since the occurrence period of different experiment conditions is segmented, here we referred to the strategy in the paper [11], by concatenating the dispersed time series belonging to the same experiment condition. Finally, we calculated the Pearson's correlation matrix of each experiment condition for each subject.

2.5 Topological Property Calculation

Network efficiency [12] is concept used in network science which is used to measure the efficiency of information exchange, and can be applied to both local and global scales. We don't understand the characteristics of the topological properties of the language network in the experiment of word and object processing. Hence, we first use the global efficiency [12, 13] to measure the information exchange efficiency at the global scale. Formula (1) is the definition of global efficiency as follows:

$$E_{global} = \frac{1}{N(N-1)} \sum_{i \neq j \in G} \frac{1}{d_{i,j}} \tag{1}$$

where G denotes the binary matrix of the functional connectivity under a specific graph density, N denotes the number of nodes in the functional brain network, $d_{i,j}$ denotes the shortest path distance between node i and node j.

The calculation of global efficiency needs to transform the Pearson's correlation matrix into a series of binary matrices at multiple graph density. The workflow of the calculation is shown as Fig. 4, we first got the correlation matrix corresponding to the functional connectivity, and then set a threshold to generate binary matrix. Since there are a certain range of coefficients in the Pearson's correlation matrix, 100 thresholds were chosen for generating the full range of graph density (1%–100%) and the increment of 1% to examine the language network under different experiment conditions of word and object processing.

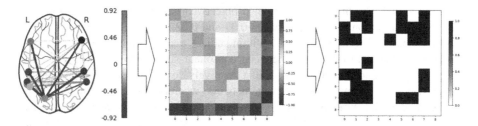

Fig. 4. Workflow of topological property calculation

3   Results and Discussions

After obtaining the global efficiencies of the language network under 100 graph densities of each experimental condition, we performed a statistical test on each pair of experiment conditions under each graph density to determine whether there is a significant difference in efficiencies of the language network. When analyzing on the network density, the sparsity of different segments has a certain difference in the importance of the results and the graph density of 0.1 to 0.5 is recommended [14]. All the statistical tested results are illustrated in Fig. 5. According to the above figure, no significant differences in the global efficiency of language network can be identified between the stimuli of consonant strings and object pictures shown in Fig. 5(a), indicating the similar information exchange efficiency of the language network under these experiment conditions. Although some differences of global efficiency of the language network can be found between the stimuli of consonant strings and scrambled objects in Fig. 5(b), and the stimuli of objects pictures and scrambled objects in Fig. 5(c). It cannot be asserted the stimulus condition under which the language network has a higher information exchange efficiency. In Fig. 5(b), there are ten graph densities under which significant differences exists between the two stimulus conditions of consonant strings and scrambled objects. The global efficiency of language network under consonant string condition is higher than that under scrambled object condition on the left five graph densities, while the opposite is true on the right five graph densities. Similarly in Fig. 5(c), there are seven graph densities under which significant differences exists between the two stimulus conditions of objects and scrambled objects. The global efficiency of language network under object condition is higher than that under scrambled object condition on the left three graph densities, while the opposite is true on the right four graph densities. Furthermore, it can be seen from Fig. 5(d) that there are eighteen graph sparsities under which the global efficiency of language network under word condition is significantly higher than that under consonant string condition, and the eighteen graph sparsities are in the range between 0.2 and 0.55. In Fig. 5(e), there are seventeen graph sparsities under which significant differences exists between stimulus conditions of words and objects. Although the global efficiency under words condition is significantly lower than that under objects condition in the two left most graph sparsities, there are still 15 graph sparsities where is the opposite and their distribution is more reasonable, that is, between 0.2 and 0.36. In Fig. 5(f), there are five graph densities under which the global efficiency of language network under words condition is significantly higher than that under scrambled objects condition, and the five graph densities are in the reasonable range between 0.09 and 0.22.

As shown in the above figures, within-subject paired t-tests demonstrated the existence of the small differences between consonant strings, objects and scrambled objects (Fig. 5(a)–(c)), which indicated a relatively lower global efficiency in the connectivity of the human brain language network, whereas quite stark differences can be identified within written words vs consonant strings, normal object picture and scrambled object picture (Fig. 5(d)–(f)), which showed that there was a higher global efficiency and a stronger connectivity of the human brain language network in the recognition task of written words compared with other task modality at the group level.

The present result could be possibly explained with the task dependent stimulus modality effects, which claims that the processing strategies of the neural network collaboration is dependent on the stimulus types. According to paper [15], the recognition of words involves the access to phonology before semantics, while the access to semantics would occur before phonology in the objects recognition. Based on the stimulus modality effects, few previous studies that have directly compared word and picture processing have produced conflicting results.

Although the general objects' recognition involves two necessary cognitive processes, including a rapid visual analysis of the object and the activation of semantic knowledge, the object recognition task in this experiment only examine the possible repetition of the same object picture. Subjects only need to take the visual strategies for pictures judgement and thus less language network nodes would be activated although some semantics related to the pictures might be automatically involved. Compared with object picture identification, written words recognition, either the true English words or the consonant strings, will start with the neuronal populations becoming tuned to word form processing, access to the phonology and then to the semantics processing. As a result, more neuro system nodes of the language network would be involved, a stronger functional connectivity would be activated and a higher global efficiency of the language network would be achieved.

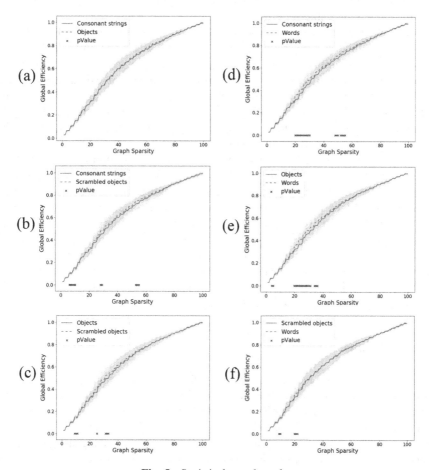

Fig. 5. Statistical tested results

To explain the differences between the recognition of true English words and that of consonant strings, it is necessary to take a further exploration into the cognitive phonological processing strategies for the written words. Compared with the true English words, the consonant strings are unpronounceable since they are not formed according to the phonological principles. In identifying the visual input of consonant strings, the human brain's language network would first focus on their phonological information. When the perceptive identification of the irregularities of the phonological codes occurs, a further access to semantics network would break up. Thus the connectivity of the language network nodes would be adversely influenced and the global efficiency of language network might be at a comparatively lower level. However, the recognition of true words involves the visual input of the words form being rapidly transformed into the phonology process. The true written words have the normal and correct phonological features, and the further access to semantics would be instantly activated after the phonological processing. Since the whole process of true written words make more language network modes get involved, a stronger functional connectivity and a higher efficiency would be achieved inevitably.

Among the four stimulus types, the picture of scrambled objects is the least visually recognizable with little information about the normal shape and color, and thus no relative semantic information could be activated. When getting the stimulus input of scrambled objects, the subjects could hardly make any judgement and identify what the object would be, even with the visual strategies involved. As a result, it would be unnecessary for the activation of further cognitive phonological and semantics processing, and the global efficiency and connectivity between the various nodes of the language network would decrease accordingly.

To sum up, the overall functional connectivity and global information exchanging efficiency of human brain language network will be hugely influenced by the stimulus modality and the different interactions between the language network nodes may cause differential activations for words and objects within the same neuronal system. At a cognitive level, written words reading, compared with other tasks, need earlier and longer activation since written words have stronger associations with phonology and the visual recognition of written words is associated with a top-down perceptual and semantic processing of signals being sent to the occipitotemporal areas. Consequently, a stronger structural connectivity and a higher information exchange efficiency will be triggered under the written words reading task condition.

4   Conclusions

The final results provide the evidence that distributed nodes of human brain language network are inter-regionally connected and activated during the mental processing tasks. The inter-communication efficiency of the language network varies as a result of the different cognitive mental processes, of which accessing the semantic memory is the key part. The communication efficiency of the human brain language network is enhanced when the cognitive task involves mental processing with the external stimuli containing semantic information (e.g. word reading), while the information exchange efficiency between the network nodes will significantly decrease in dealing with the external stimuli with less semantic information (e.g. the picture of scrambled object). Furthermore, no significant differences between the communication efficiencies of the language network have been identified in the task performance of object recognition, scrambled object recognition and consonant strings reading, despite that the possibly different semantic processing might get involved. Clearly, the precise differences of language network communication efficiency between the similar processes in semantics remain to be specified. In particular, the idea that how and to what extend the inter-regional connectivity and global information exchanging efficiency of human language network are influenced by different cognitive task types awaits further study.

Acknowledgements. This work was supported by grants from the National Natural Science Foundation of China (61420106005), Ministry of Education Humanities & Social Sciences of China (14YJC740030), the Science and Technology Project of Beijing Municipal Commission of Education (KM201710005026), and the JSPS Grants-in-Aid for Scientific Research of Japan (19K12123).

References

1. Downing, P.E., Chan, A.W., Peelen, M.V., et al.: Domain specificity in visual cortex. Cereb. Cortex **16**(10), 1453–1461 (2006)
2. Jiang, X., Bradley, E., Rini, R.A., et al.: Categorization training results in shape- and category-selective human neural plasticity. Neuron **53**(6), 891–903 (2007)
3. Spiridon, M., Fischl, B., Kanwisher, N.: Location and spatial profile of category-specific regions in human extrastriate cortex. Hum. Brain Mapp. **27**(1), 77–89 (2010)
4. Von, K.K., Dogan, O., Grüter, M., et al.: Simulation of talking faces in the human brain improves auditory speech recognition. Proc. Natl. Acad. Sci. U.S.A. **105**(18), 6747–6752 (2008)
5. Yovel, G., Tambini, A., Brandman, T.: The asymmetry of the fusiform face area is a stable individual characteristic that underlies the left-visual-field superiority for faces. Neuropsychologia **46**(13), 3061–3068 (2008)
6. Rubinov, M., Sporns, O.: Complex network measures of brain connectivity: uses and interpretations. NeuroImage **52**(3), 1059–1069 (2010)
7. Duncan, K.J., Pattamadilok, C., Knierim, I., et al.: Consistency and variability in functional localisers. Neuroimage **46**(4), 1018–1026 (2009)
8. Rees, G.: Statistical Parametric Mapping. Science Press, China (2010)
9. Abraham, A., Pedregosa, F., Eickenberg, M., et al.: Machine learning for neuroimaging with scikit-learn. Front. Neuroinform. **8**(14), 14 (2013)
10. Richiardi, J., Altmann, A., Milazzo, A.C., et al.: Correlated gene expression supports synchronous activity in brain networks. Science **348**(6240), 1241–1244 (2015)
11. Cohen, J.R., D'Esposito, M.: The segregation and integration of distinct brain networks and their relationship to cognition. J. Neurosci. Off. J. Soc. Neurosci. **36**(48), 12083–12094 (2016)
12. Latora, V., Marchiori, M.: Efficient behavior of small-world networks. Phys. Rev. Lett. **87**(19), 1–4 (2001)
13. Latora, V., Marchiori, M.: Economic small-world behavior in weighted networks. Eur. Phys. J. B-Condens. Matter Complex Syst. **32**(2), 249–263 (2003)
14. Watts, D.J., Strogatz, S.H.: Collective dynamics of 'small-world' networks. Nature **393**(4), 440–442 (1998)
15. Glaser, W.R., Glaser, M.O.: Context effects in stroop-like word and picture processing. J. Exp. Psychol. Gen. **118**(1), 13–42 (1989)

The Neural Mechanism of Working Memory Training Improving Emotion Regulation

Xiaobo Wang[1,2,3], Dongni Pan[1,2], and Xuebing Li[1,2(✉)]

[1] Institute of Psychology, Chinese Academy of Sciences, Beijing 100101, China
lixb@psych.ac.cn
[2] Department of Psychology, University of Chinese Academy of Sciences,
Beijing 100049, China
[3] North China University of Technology, Beijing 100144, China

Abstract. Thirty-six patients with high anxiety were recruited. The subjects were divided into working memory training group and control group in a voluntary and random manner, with 18 individuals in each group. The training group was trained for 21 days of working memory, and the control group was not trained for working memory. The subjective emotion ratings and the ERP indicator late positive potential (LPP) of the two groups of participants were recorded, under three experimental conditions (watching negative images, cognitive reappraisal, attentional distraction). It was found that the LPP amplitude reduction was significantly higher for training group than control group specifically in the condition of cognitive reappraisal. This study showed that working memory training can improve the ability cognitive reappraisal and can be a potential intervention for promoting the emotional regulation of individuals with high trait anxiety.

Keywords: Memory training · Emotional regulation · Neural mechanism · Dual N-back · Anxiety

1 Introduction

Working memory is the ability to temporarily maintain and manipulate information as individuals perform cognitive tasks [1]. It plays an important role in human cognition. Emotion regulation affects deeply in our mental health and the emotional regulation strategies (including cognitive reappraisal and distraction [2, 3] have a certain correlation with life satisfaction, positive emotions, depression, anxiety.

Some studies have established that working memory ability and emotion regulation are connected. Working memory capacity could control an individual's attention, which is very important for emotional regulation [4]. Some scholars had found that the emotional regulation framework of selection, optimization and compensation required internal resources, and this required people's ability to control attention and working memory [5]. Other researchers directly reported that in a down-regulation task,

This research was supported by National Nature Science Foundation of China [NSFC 31671136, 31530031, 51708003].

P. Liang et al. (Eds.): BI 2019, LNAI 11976, pp. 72–81, 2019.
https://doi.org/10.1007/978-3-030-37078-7_8

participants with higher working memory capacity experienced and expressed fewer emotional responses; and the participants with lower working memory capacity were more susceptible to emotional contagion and less successful in applying reappraisal strategies [6].

Previous study had found that working memory training could improve individuals' working memory capacity [7], and could affect other cognitive functions associated with working memory, such as various executive functions and attention control [8–10]. The effect extended even to the emotional realm. Studies had found that extensive working memory training could weaken the anger, exhaustion and depression of the participants [11]. and improved the heart rate variability (HRV), which could reflect the ability of emotional regulation [12].

In recent years, more and more researches focused on the relationship between emotional regulation and anxiety. A meta-analysis found that infrequent use of appropriate emotional regulation strategies leaded to more anxiety [13]. Neurological studies related to emotional regulation found that poor regulation was associated with higher levels of anxiety [14, 15].

A previous study showed that working memory training could reduce the anxiety level of anxiety patients and healthy individuals. Improvement of working memory ability could improve the performance of high-anxiety individuals in various cognitive tasks [16]. Individuals who had working memory training with high-task participation, had a significant decline in their trait anxiety compared to before [17]. But it's not known whether working memory training improves emotional regulation in individuals with high anxiety, although it has been shown in healthy people. Cognitive reappraisal and attentional distraction are commonly used emotional regulation strategies. Compared with distraction, reappraisal is considered to be a more adaptive emotional regulation strategy and was considered to be helpful to relieve anxiety [18, 19].

ERP is a special brain evoked potential with excellent temporal resolution and accurate capture of rapid emotional responses. It is ideal for studying the temporal processing of emotions. Previous ERP studies of emotional regulation had found that late positive potential (LPP) was a good electrophysiological indicator of emotional regulation [20]. The LPP amplitude of the negative emotion pictures is significantly larger than the neutral ones [21]. Besides, LPP amplitude decreases during emotion regulation [22], and thus LPP could be used as an emotion regulation index [23–25].

Based on the existing research, this study would like to explore the effects of working memory training on emotional regulation in high-anxiety individuals, using both behavior and neural indictors The training tasks was a dual n-back working memory task based on smartphone APP. And the regulation included two strategies of reappraisal and distraction. We hypothesized that working memory training could improve the adaptive emotional regulation i.e., cognitive reappraisal ability of individuals with high anxiety, which was shown in the decreased LPP during reappraisal in training group.

2 Method

2.1 Participants

Participants were selected by means of the trait sub-scale of State Trait Anxiety Inventory (STAI-T) A total of 36 individuals with STAI-T score higher than 50 were recruited. All the participants were college students and graduate students.

Participants were randomly assigned to training group18 (5 males and 13 females) and control group 18 (7 males, 11 females). The age distribution of the subjects was 18–26 years old, with an average age of 21.47 ± 2.408. Gender differences between the different groups were not significant, $\chi^2(1) = 0.50$, p = 0.48; Age differences were not significant, F(1,34) = 2.654, p = 0.113.

All participants signed an informed consent form and received a certain amount of compensation. The study was approved by the Ethics Committee of the Institute of Psychology of the Chinese Academy of Sciences.

2.2 The Task of Working Memory Training

The Task of Training Group

The task of the experimental group in this study was working memory training, which used a dual n-back task. The dual n-back task used the feature attributes of color and position. This task was nested in a self-developed Android app for mobile devices with Android 5.0 and above [26].

The training material was a nine-square grid with a color block. The training task was to ask the subjects to remember the color (one of the four colors of red, yellow, green and blue) and the position (3 × 3, 9 positions) of the color block appearing in one of the nine squares. Participants needed to determine whether the color and position of the current patch were the same as the color and position of the nth patch forward. If they were the same, the corresponding button according to the color or position at the bottom of the screen should be pressed; if they were different, the screen button should not be pressed.

In the n-back task, the number of levels was the number of n. Level 1, the participant only needed to compare whether the color and position of the current color block were the same as the previous one. Level 2, the subjects needed to compare whether the current color block color and position were the same as the second one before, and so on. The total number of trials per group was 20 + n. In each trial, the color block was shown for 500 s and the blank screen was 2500 ms. The target test of color and position was random within 4–6 ranges to reduce the use of the guessing strategy of the subjects. Each subject needed 30 training sessions per day, about 30 min. Participants were required to complete 21 days of training within 30 days. These 21 days can be continuous or dispersed.

When the subjects training, the APP's interface would feedback the following information immediately: hit, missed or missed. So that the subjects could quickly become familiar with the rules and improve the training motivation. After the training, the APP interface could feedback the number of hits, missed hits, misses, and total score of the training. Total score = (hit number/target number) × (correct rejection number/non-target number) × 100%. The total score was (0–100). This APP adopted an adaptive paradigm, because adaptive APP was considered to be a necessary condition for improving training level and obtaining far transfer effects [27]. If the scores of the two consecutive groups are >80, the APP would automatically enter the next level; if the test scores were lower than 10 for two consecutive groups, then level would return to the previous level. Daily training would clear the historical scores, that was, the participants would be trained from the first level on each day.

The Task of Control Group
In order to eliminate the possibility that the APP working memory training task might be effective as a placebo, the subjects in the control group were also arranged doing something like a task. As an active control group, they received tweets of psychology and test fee twice a week without any working memory requirements.

2.3   The Task of ERP Emotional Regulation

There were three condition during emotional regulation task: watching negative emotional images, watching negative emotional images with cognitive reappraisal, and watching negative emotional images with attentional dispersion.

The first one was watching negative emotion images, without active emotional regulation strategies, which we called watching negative images condition (WNIC).

The second one was watching negative emotion images meanwhile using cognitive reappraisal, with active emotional regulation strategies, which we called cognitive reappraisal condition (CRC). Reappraisal was a cognitive-verbal strategy that alters the path of emotional responses by reconstructing the meaning of a situation.

The third one was watching negative emotion images meanwhile using attentional dispersion, with active emotional regulation strategies, which we called attentional dispersion condition (ADC).

According to the previous research [28], the guidance for WNIC was "Then you will see some pictures. Please watch each picture carefully and respond naturally." The guidance for CRC was "Then you will see some pictures. Please watch each picture. At the same time, we hope that you can review the picture with an analytical eye, just like watching a movie, or imagine that the picture is artificially synthesized by photoshop." The guidance for ADC was "Then you will see some pictures. When the picture appears, please pay attention to the non-emotional part of the picture, such as the hair, not the features of the face." The order in which the three conditional blocks appear was random for each subject. These tasks had exercises and were guided by main tests.

In the each viewing or regulation trial, a fixed point appeared first for 500 ms, followed by the instruction text of a certain condition for 1000 ms. Each negative picture was presented for 5000 ms. Participants needed to apply an emotional regulation strategy consistent with the instructions while watching the picture. Than the subjects needed to complete the 9-level score of emotional valence and arousal, valence 1 (very negative) to 9 (very positive); arousal 1 (very weak)) to 9 (very strong) for this picture. There were 180 negative pictures (560 × 420 pixels) from the International Affective Picture System [29]. 90 for the pretest, 90 for the posttest. The images had no significant difference in valence and arousal between the images the pretest and the posttest. Every 90 pictures were randomly assigned to the 3 conditions, 30 for WNIC, 30 for CRC, 30 for ADC. There was no significant difference in valence and arousal among the 3 conditions.

The subjects sat in the soundproofing laboratory to watch the pictures. Images were presented in the center of the screen using E-Prime 2.0 presentation software. The line of sight was 60 cm and the viewing angle was 9.46 × 7.13°.

2.4   Procedures

After the successful recruitment of the training group and the control group, the participants entered the laboratory to receive the emotion regulation ERP task. Then the training group was trained for 21 days of working memory, and the control group did not perform working memory training. After 30 days, all subjects re-entered the laboratory and received the emotion regulation task again. Experimental flow chart was as follows (see Fig. 1).

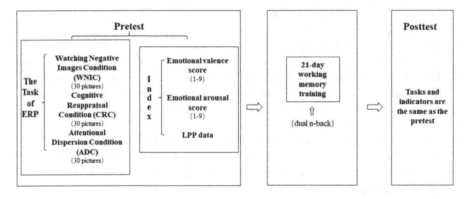

Fig. 1. Experimental flow chart

2.5   EEG Recordings

ERP can be used as a sensitive indicator of emotion. We chose late positive potential (LPP) and a time window of 350–800 ms to assess the impact of work memory training tasks on emotional regulation. 64 electrodes placed on an elastic scalp (Neuro Scan 4.5) were used to record the electrical activity. According to the distribution of the topographic map, we selected the electrodes of P5, P3, P1, PZ, P2, P4, P6, P8, PO5, PO3, POZ, PO4, PO6. In order to reduce the type 1 error, we averaged the data of these electrode points. All electrode impedances were maintained below 5 KΩ. EEG was sampled at the rate of 1000 Hz, by a 0.05- to 100-Hz band-pass amplifying. A low-pass filter at 30 Hz (12 dB/oct) were used to filter the EEG data. There was a baseline correction of 500 ms pre stimulus. According to the images' onset in the emotion regulation task, ERP data were segmented for each condition from −500 ms to 2,000 ms. Trials containing activity over ±80 μV were excluded from averaging. Each block had at least 25 effective trials.

2.6   Data Analysis

All these statistical analyses were conducted with SPSS Version 20.0 software. Firstly, the study analyzed the variance analysis with repeated measures for group (2) × condition (3) × time (2), and analyzed the main effect and interaction effect Then we conducted independent sample t-test for the difference between the pre-test and post-test of the 3 conditions, to determine whether there was significant difference between groups or conditions.

3   Result

3.1   Emotional Valence and Arousal

Valence
Repeated measures analysis of variance for the valence score showed that the conditional main effect was significant, $F(2,68) = 151.5$, $p = 0.00$, $\eta p2 = 0.817$. Post hoc comparisons showed the scores of WNIC (M = 3.338, SE = 0.100) were significantly lower than the scores of CRC (M = 4.335, SE = 0.089), and the scores of ADC (M = 4.716, SE = 0.079). $p < 0.05$. Different conditions and different groups, the average value of the test before and after was shown in the Fig. 2.

The main effects of the groups were not significant $F(1,34) = 0.748$, $p = 0.393$, $\eta p2 = 0.022$, and the interaction between the groups and conditions was not significant $F(1,34) = 2.887$, $p = 0.098$, $\eta p2 = 0.078$.

Arousal
Repeated measures analysis of variance for the arousal score showed that the conditional main effect was significant, $F(2,68) = 55.398$, $p < 0.001$. Post hoc

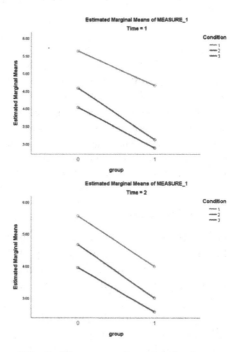

Fig. 2. The means of emotional valance

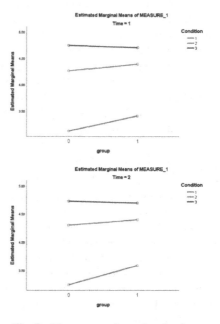

Fig. 3. The means of emotional valance

comparisons showed the scores of WNIC(M = 4.959,SE = 0.215) were significantly higher than the scores of CRC (M = 3.842, SE = 0.212), and the scores of ADC (M = 3.357, SE = 0.209), p < 0.05. Different conditions and different groups, the average value of the test before and after was shown in the Fig. 3.

The control group (M = 4.737, SE = 0.271) was significantly higher than the training group (M = 3.369, SE = 0.271). The interaction between the group and the condition was not significant, F(2,68) = 0.576, p = 0.565. The interaction between the group and time was not significant, F(2,68) = 0.823, p = 0.371.

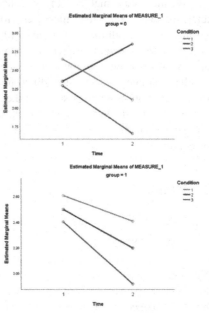

3.2   LPP Indicator

For the group (2) × condition (3) × time (2), the variance analysis with repeated measures was performed. The main effect of condition was significant, F = 3.922, p = 0.03, ηp2 = 0.192.

Post hoc comparisons showed that the LPP scores were as follows: WNIC (M = 2.441, SE = 0.197), CRC (M = 2.472, SE = 0.210), ADC (M = 2.065, SE = 0.184). The negative images were significantly higher than the attentional dispersion (p = 0.017), and the cognitive reappraisal was signifi-cantly higher than the attentional dispersion (p = 0.013).

Fig. 4. The means of LPP

The main effect of time was marginal significant, F = 3.944, p = 0.055, ηp2 = 0.104. Pairwise comparisons showed that the pretest (M = 2.465, SE = 0.209) was greater than the posttest (M = 2.187, SE = 0.173).

The interaction of time and condition was significant, F = 5.199, p = 0.11, ηp2 = 0.240. The triple interaction of group, condition and time was significant, F = 4.440, p = 0.020, ηp2 = 0.212.

The LPP differences between pretest and posttest of WNIC, CRC and ADC were calculated. The difference = Pretest-Posttest. This study compared the difference of the 3 conditions. The results showed that only in the condition of cognitive reappraisal, the difference of LPP of the training group reached a significant level, t(34) = −2.082, p = 0.045. But, this was not observed significantly in the other two conditions (watching negative images and attentional dispersion).

The estimated marginal average of the control and training groups was shown in Fig. 4. The figure showed, for the working memory training group, the LPP value of the post-test was decreased regardless of the pre-test situation. The low LPP value reflected the low response to negative images. This showed that after working memory training, the ability to adjust emotions had become better.

4 Discussion

The current study examined the effect of working memory training on emotional regulation in anxious individuals. For the different emotional regulation strategies of high anxiety individuals, both cognitive reappraisal and attention dispersion are effective.

But, statistical analysis of LPP data showed that working memory training specifically enhanced the ability of individual cognitive reappraisal. Perhaps because cognitive reappraisal requires more participation of working memory, the working memory training can have more impact on this strategy of emotional regulation.

Analysis the neural mechanism of working memory improving the emotional regulation, we believe that working memory requires the participation of the prefrontal lobe. Through the training of working memory, the prefrontal lobe control ability is enhanced, thus changing the cognitive reappraisal of anxious individuals. The results of this study are consistent with the conclusions of an existing study that suggested that the refresh function of working memory may migrate to the trainee's cognitive reappraisal ability, thereby improving the trainee's emotional regulation ability [30].

Further analysis found that the theoretical structure of working memory and emotional regulation have certain similarities. The core function of working memory is to store and process information. In addition to the short-term preservation of information, it also involves processing, such as updating, suppressing, and shifting. The process of emotional regulation is similarities. Emotional regulation requires suppression of bad emotional information, while constantly updating and transforming emotional value information. Therefore, the working memory process can be seen as the cognitive basis of emotional regulation.

How people use emotional regulation strategies to regulate emotions is important for understanding the causes, maintenance, and treatment of anxiety. We found that working memory training can improve the emotional regulation of anxious individuals and promote the use of adaptive cognitive reappraisal, so it is a potential intervention.

However, the number of participants is small, and only subclinical individuals with high anxiety were examined. Future research needs to further investigate the application of training in clinical groups to prove the authenticity of emotional benefits of working memory training.

5 Conclusion

Working memory training can significantly reduce the LPP of cognitive reappraisal in high anxiety individuals, and specifically improve the effectiveness of individual's cognitive reappraisal ability, indicating a promising intervention for anxious people.

References

1. Baddeley, A.: Working memory. Science **255**(5044), 556–559 (1992)
2. Goldin, P.R., McRae, K., Ramel, W., Gross, J.J.: The neural bases of emotion regulation: reappraisal and suppression of negative emotion. Biol. Psychiatry **63**(6), 577–586 (2008)

3. Hermann, A., Bieber, A., Keck, T., Vaitl, D., Stark, R.: Brain structural basis of cognitive reappraisal and expressive suppression. Soc. Cogn. Affect. Neurosci. **9**(9), 1435–1442 (2013)
4. Wadlinger, H.A., Isaacowitz, D.M.: Fixing our focus: training attention to regulate emotion. Pers. Soc. Psychol. Rev. **15**(1), 75–102 (2011)
5. Opitz, P.C., Gross, J.J., Urry, H.L.: Selection, optimization, and compensation in the domain of emotion regulation: applications to adolescence, older age, and major depressive disorder. Soc. Pers. Psychol. Compass **6**(2), 142–155 (2012)
6. Schmeichel, B.J., Volokhov, R.N., Demaree, H.A.: Working memory capacity and the self-regulation of emotional expression and experience. J. Pers. Soc. Psychol. **95**(6), 1526 (2008)
7. Klingberg, T., Forssberg, H., Westerberg, H.: Training of working memory in children with ADHD. J. Clin. Exp. Neuropsychol. **24**(6), 781–791 (2002)
8. Dahlin, E., Nyberg, L., Bäckman, L., Neely, A.S.: Plasticity of executive functioning in young and older adults: immediate training gains, transfer, and long-term maintenance. Psychol. Aging **23**(4), 720 (2008)
9. Zhao, X., Zhou, R., Fu, L.: Working memory updating function training influenced brain activity. PLoS One **8**(8), e71063 (2013)
10. Clark, C.M., Lawlor-Savage, L., Goghari, V.M.: Functional brain activation associated with working memory training and transfer. Behav. Brain Res. **334**, 34–49 (2017)
11. Takeuchi, H., et al.: Working memory training improves emotional states of healthy individuals. Front. Syst. Neurosci. **8**, 200 (2014)
12. Xiu, L., Zhou, R., Jiang, Y.: Working memory training improves emotion regulation ability: evidence from HRV. Physiol. Behav. **155**, 25–29 (2016)
13. Schäfer, J.Ö., Naumann, E., Holmes, E.A., Tuschen-Caffier, B., Samson, A.C.: Emotion regulation strategies in depressive and anxiety symptoms in youth: a meta-analytic review. J. Youth Adolesc. **46**(2), 261–276 (2017)
14. DeCicco, J.M., Solomon, B., Dennis, T.A.: Neural correlates of cognitive reappraisal in children: an ERP study. Dev. Cogn. Neurosci. **2**(1), 70–80 (2012)
15. Dennis, T.A., Hajcak, G.: The late positive potential: a neurophysiological marker for emotion regulation in children. J. Child Psychol. Psychiatry **50**(11), 1373–1383 (2009)
16. Qi, S., et al.: Impact of working memory load on cognitive control in trait anxiety: an ERP study. PloS One **9**(11), e111791 (2014)
17. Sari, B.A., Koster, E.H., Pourtois, G., Derakshan, N.: Training working memory to improve attentional control in anxiety: a proof-of-principle study using behavioral and electrophysiological measures. Biol. Psychol. **121**, 203–212 (2016)
18. Aldao, A., Nolen-Hoeksema, S., Schweizer, S.: Emotion-regulation strategies across psychopathology: a meta-analytic review. Clin. Psychol. Rev. **30**(2), 217–237 (2010)
19. Hu, T., Zhang, D., Wang, J., Mistry, R., Ran, G., Wang, X.: Relation between emotion regulation and mental health: a meta-analysis review. Psychol. Rep. **114**(2), 341–362 (2014)
20. Mocaiber, I., et al.: Fact or fiction? An event-related potential study of implicit emotion regulation. Neurosci. Lett. **476**(2), 84–88 (2010)
21. Hajcak, G., Nieuwenhuis, S.: Reappraisal modulates the electrocortical response to unpleasant pictures. Cogn. Affect. Behav. Neurosci. **6**(4), 291–297 (2006)
22. Hajcak, G., MacNamara, A., Olvet, D.M.: Event-related potentials, emotion, and emotion regulation: an integrative review. Dev. Neuropsychol. **35**(2), 129–155 (2010)
23. Parvaz, M.A., Moeller, S.J., Goldstein, R.Z., Proudfit, G.H.: Electrocortical evidence of increased post-reappraisal neural reactivity and its link to depressive symptoms. Soc. Cogn. Affect. Neurosci. **10**(1), 78–84 (2014)
24. Frühholz, S., Jellinghaus, A., Herrmann, M.: Time course of implicit processing and explicit processing of emotional faces and emotional words. Biol. Psychol. **87**(2), 265–274 (2011)

25. Hajcak, G., Dunning, J.P., Foti, D.: Motivated and controlled attention to emotion: time-course of the late positive potential. Clin. Neurophysiol. **120**(3), 505–510 (2009)
26. Pan, D.N., Wang, D., Li, X.-B.: Cognitive and emotional benefits of emotional dual dimension n-back training based on an APP. Acta Psychologica Sinica **50**(10), 1105–1119 (2018)
27. Pedullà, L., et al.: Adaptive vs. non-adaptive cognitive training by means of a personalized App: a randomized trial in people with multiple sclerosis. J. NeuroEng. Rehabil. **13**(1), 88 (2016)
28. Pan, D.N., Wang, Y., Li, X.: Strategy bias in the emotion regulation of high trait anxiety individuals: an investigation of underlying neural signatures using ERPs. Neuropsychology **33**, 111–122 (2019)
29. Lang, P.J., Bradley, M.M., Cuthbert, B.N.: International affective picture system (IAPS): instruction manual and affective ratings. The Center for Research in Psychophysiology, University of Florida (1999)
30. Pe, M.L., Raes, F., Kuppens, P.: The cognitive building blocks of emotion regulation: ability to update working memory moderates the efficacy of rumination and reappraisal on emotion. PloS One **8**(7), e69071 (2013)

Dynamic Functional Connectivity in the Musical Brain

Dipankar Niranjan[1]([✉]), Petri Toiviainen[2], Elvira Brattico[3], and Vinoo Alluri[1]

[1] Cognitive Science Lab, Kohli Center on Intelligent Systems,
IIIT Hyderabad, Hyderabad, India
dipankar.niranjan@research.iiit.ac.in, vinoo.alluri@iiit.ac.in
[2] Department of Music, University of Jyvaskyla, Jyväskylä, Finland
petri.toiviainen@jyu.fi
[3] Department of Clinical Medicine, Center for Music in the Brain,
Aarhus University, Aarhus, Denmark
elvira.brattico@clin.au.dk

Abstract. Musical training causes structural and functional changes in the brain due to its sensory-motor demands. This leads to differences in how musicians perceive and process music as compared to non-musicians, thereby providing insights into brain adaptations and plasticity. Correlational studies and network analysis investigations have indicated the presence of large-scale brain networks involved in the processing of music and have highlighted differences between musicians and non-musicians. However, studies on functional connectivity in the brain during music listening tasks have thus far focused solely on static network analysis. Dynamic Functional Connectivity (DFC) studies have lately been found useful in unearthing meaningful, time-varying functional connectivity information in both resting-state and task-based experimental settings. In this study, we examine DFC in the fMRI obtained from two groups of participants, 18 musicians and 18 non-musicians, while they listened to a musical stimulus in a naturalistic setting. We utilize spatial Group Independent Component Analysis (ICA), sliding time window correlations, and a deterministic agglomerative clustering of windowed correlation matrices to identify quasi-stable Functional Connectivity (FC) states in the two groups. To compute cluster centroids that represent FC states, we devise and present a method that primarily utilizes windowed correlation matrices occurring repeatedly over time and across participants, while excluding matrices corresponding to spontaneous fluctuations. Preliminary analysis indicate states with greater visuo-sensorimotor integration in musicians, larger presence of DMN states in non-musicians, and variability in states found in musicians due to differences in training and prior experiences.

Keywords: Dynamic Functional Connectivity · Clustering · ICA · State characterization · Musicians vs. non-musicians

P. Liang et al. (Eds.): BI 2019, LNAI 11976, pp. 82–91, 2019.
https://doi.org/10.1007/978-3-030-37078-7_9

1   Introduction

Professional musicians typically undergo an intensive formal training period that lasts several years. The training is followed by consistent practice and performance, often running into several hours per week. This intensive sensory-motor training causes structural [11,13] and functional [4,10] changes in the brain. Musicians also have different cerebral characteristics which correlate with the age of commencement of training and also the intensity/frequency of training [14]. This makes music a great tool to study brain adaptation, and musicians an ideal group to study brain changes driven by experience, especially when contrasted with a non-musicians group.

Moreover, since music is inherently multidimensional in nature, there has been an increased focus on the use of naturalistic stimulus in continuous music listening settings (emulating real-life listening experiences). These investigations have been shown to present a more holistic picture of the neural underpinnings of music processing, as against those performed in controlled auditory settings where musical features are often presented in isolation and manipulated artificially. Correlational studies following this paradigm have indicated the presence of large-scale brain networks (involving the recruitment of cognitive areas of the cerebellum, sensory and DMN cerebrocortical areas, and motor and emotion-related circuits), in musicians, involved in the processing of musical features like timbre, rhythm and tonality [3,7,20]. Furthermore, network studies have also been conducted to highlight functional networks and key hubs recruited during music listening [15,21]. In a study which is more in line with our work, static whole-brain functional connectivity analyses revealed group-differences between musicians and non-musicians, with the primary hubs of the musicians consisting of the cerebral and cerebellar sensorimotor regions, and those of the non-musicians consisting of DMN-related regions [2]. Community structure analyses of the key hubs revealed greater integration of motor and somatosensory homunculi representing the upper limbs and torso in musicians. Network investigations in this domain have thus far been restricted to static Functional Connectivity (FC) analysis.

Thus it follows that the assessment of FC in these studies has largely been limited by an assumption of spatial and temporal stationarity throughout the fMRI scan period. While this presents a simple template for static whole brain connectivity analysis, it comes at the cost of an inability to study FC patterns across scan timecourses. To enable dynamic temporal analysis, researchers have suggested various methods to identify and characterize FC states leading to interesting findings in FC patterns over time, in task-based and resting-state analyses [6,12,18]. In this study, we utilize and extend the theoretical model and framework proposed by Allen et al. [1] on fMRI data obtained in a task-free, continuous music listening setting.

We begin by identifying Intrinsic Connectivity Networks (ICNs) using a group-level (musicians and non-musicians) ICA analysis on the fMRI data. We perform sliding window correlation computations on the time-courses of the back-reconstructed ICNs. Finally, to identify quasi-stable states which repeat

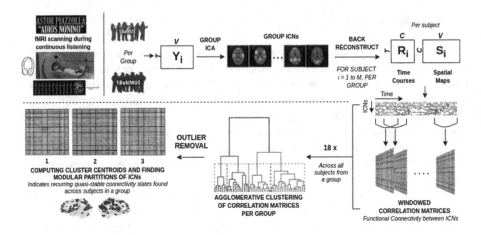

Fig. 1. An overview of our study.

across participants in the group, we adopt an agglomerative clustering approach to cluster windowed correlation matrices across all participants of the group. In earlier work [1,9], all of the matrices were used in the computation of centroids. Deviating from this, we hypothesize that FC states are of two types: ones which recur over time and across participants, and ones which are reflective of spontaneous fluctuations that do not represent generalizable group characteristics. To account for this, we include a step to identify and select matrices which repeat over time and occur across participants, and exclude outliers corresponding to subject specific activations and spontaneous fluctuations. We then find community structures in these FC states through the Louvain modularity-maximization method [5].

2   Methods

2.1   Participants, Stimulus, fMRI Data Acquisition

The participant pool consisted of 18 musically trained (9 female, mean age: 28) and 18 untrained (10 female, mean age: 29) participants. Both groups were comparable with respect to cognitive measures (WAIS-WMS III scores) and socioeconomic status (Hollingshead's FFI). The total number of years of training for musicians was 16 ± 5.7 years. The number of hours spent practicing music on average per week was 16.6 ± 11. The data was collected as part of a broader project ("Tunteet") involving other tests (neuroimaging and neurophysiological measures). The study protocol was approved by the ethics committee of the Coordinating Board of the Helsinki and Uusimaa Hospital District. Written consent was obtained from all the participants. They were asked to listen to an instrumental nuevo tango piece - Adios Nonino by Astor Piazzolla. This piece consisted of a high amount of variation in acoustic features such as timbre,

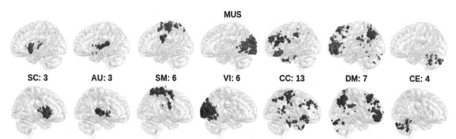

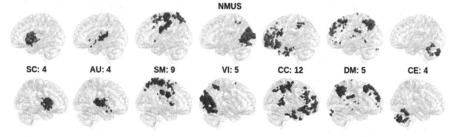

Fig. 2. Lateral views of the left (at top) and right (at bottom) hemispheres for both the groups indicating the spatial maps of the ICNs grouped as indicated (views show the union of the ICNs for each grouping). The numbers indicate the number of selected ICNs which belong to that group. Most of the regions are common to both the Mus and NMus groups and are indicated in the middle.

tonality, rhythm etc., and was ~8 min in duration. Participants' brain responses were acquired while they listened to the musical stimulus. Their only task was to attentively listen to the music delivered via MR-compatible insert earphones. MRI data was collected at the Advanced Magnetic Imaging Centre, Aalto University, Finland, on a 3T Siemens Skyra, TR = 2 s, TE = 32 ms, whole brain, voxel size: $2 \times 2 \times 2 \text{ mm}^3$, 33 slices, FoV: 192 mm (64×64 matrix), interslice skip = 0 mm. fMRI scans were preprocessed on Matlab using SPM8, VBM5 and custom scripts. Normalization to MNI segmented tissue template was carried out. Head movement related components were regressed out, followed by spline interpolation and filtering. Then, the voxel time series was Z-scored.

2.2   Group ICA and Postprocessing

Functional data from both the groups were separately analyzed using spatial Group ICA (GICA) implemented in the GIFT toolbox [8]. We chose not to group data from both groups before GICA as that would result in reduced sensitivity to between-group differences [16]. A subject-level PCA step was first used to reduce 232 time point data (464 s of music at TR = 2 s) into 180 dimensions. This data was concatenated across time (over subjects) and a group PCA step reduced this stacked matrix into 100 components. 100 independent components (aggregated across 10 runs) were obtained from the group PCA reduced matrix using the Infomax algorithm. Per participant spatial maps (SMs) and time courses (TCs) were obtained using the spatiotemporal regression back reconstruction approach [8]. Per participant SMs and TCs underwent post-processing as described in [1]. We normalized the variance of each TC, thus covariance matrices (below) correspond to correlation matrices. ICNs were identified using thresholded one sample t-test maps resulting in $C_{mus} = 42$ and $C_{nmus} = 43$ ICNs (stability index Iq > 0.9) chosen out of the 100 independent components. The ICNs were segregated into 7 groups (based on their anatomical and functional properties) indicative of Subcortical (SC), Auditory (AU), Sensorimotor (SM), Visual (VI), Cognitive Control (CC), Default Mode Network (DM), and Cerebellar (CE) regions as shown in Fig. 2.

2.3   DFC and Clustering

As in Allen et al., and Damaraju et al. [1,9], for each subject $i = 1 \dots N$, we estimate Dynamic FC using a sliding window approach, where covariance matrices are computed from windowed segments of R_i (Fig. 1). We utilize a tapered rectangular sliding window (Fig. 1) of 30 TRs, slid in steps of 1 TR, and convolved with a Gaussian of $\sigma = 3$ TRs (to obtain edge tapering), resulting in 202 windows per participant. Covariance was estimated from the regularized inverse covariance matrix (ICOV) using the graphical LASSO framework. An additional L1 norm constraint was imposed on the covariance matrix to enforce sparsity. The regularization parameter was optimized for each subject by evaluating the log-likelihood of unseen data of the subject in a cross-validation framework. After computing DFC values for each subject, covariance (correlation) values were Fisher-Z transformed to stabilize variance. As mentioned earlier, since the TCs were variance normalized, these covariance matrices correspond to correlation matrices and will henceforth be referred to as such.

As many DFC patterns recur within subjects across time, and also occur across subjects, we performed a group level (musicians and non-musicians) clustering analysis to identify the states represented by these recurring patterns. Per group, we cluster all the 3636 (18 subjects × 202 matrices = 3636) correlation matrices computed earlier. We deviate from prior work with regard to the clustering technique and the distance metric used. We chose to adopt Agglomerative Clustering with complete linkage, using cosine distance. Agglomerative Clustering is a deterministic method which has an added advantage of being able to provide a dendrogram to visualize cluster spreads and the hierarchy leading to the formation of FC states. Euclidean distance metrics do not lend themselves well in a sparse, high dimensional setting [22] (in our setting, we have 3636 vectors with each being 903 (861 for NMUS) dimensional ($\binom{43}{2} = 903$)). A cosine distance metric is better suited in such cases. We also wanted to capture similar states across the entire group, and not long chains of FC windows from individual participants (the chosen time step of 1TR leads to high autocorrelation in the FC timeseries). Hence we utilized complete linkage as against single/average linkage (Ward's method is ruled out due to our choice of a cosine distance metric). The optimal number of clusters (k) was determined using the standard elbow criterion - plotting sum of squared errors at each k, using a cosine distance metric. We also validate our choice by visually inspecting the clustering using the dendrogram. At an optimal k, splitting the dendrogram at a lower position (greater k value corresponding to more number of clusters) in the tree, as against splitting it at k, gives rise to two (or more) similar disjoint centroid states which occur at the same *level* in the tree (when ideally they should be a part of the same cluster). Using this method we find 4 clusters in the musicians and 3 in the non-musicians.

Each of these k clusters is composed of two broad types of correlation matrices - those which recur over time and across participants (these are precisely the matrices which are indicative of quasi-stable FC states and should be included in centroid computation) and those which correspond to subject specific activations and spontaneous fluctuations (which should ideally be excluded from centroid computation). We perform the following steps:

1. At this stage, in each group, the windowed correlation matrix timeseries of each subject is composed of strips of contiguous matrices belonging to one of the k clusters. For subject i, we denote the 202 matrices as $m_{i1}, m_{i2}, \dots m_{i202}$. We consider strips with atleast 10 (chosen empirically) contiguous correlation matrices belonging to the same cluster and denote them per subject as $s_{i1} \dots s_{ij}$ when $j = 1 .. J$ such strips exist for participant i. We denote the median correlation matrix (ordered by time) for each such s_{ij} as med_{ij}. For eg: $m_{3\ 17} - m_{3\ 37}$ could belong to cluster $k = 2$ for subject 3 in the musicians group. The median matrix med_{3j} for strip s_{3j} would be $m_{3\ 27}$.

2. To get a better estimate of the cluster center for cluster k, we choose one median matrix per subject, such that the pairwise-sum of the cosine distances between the chosen matrices is minimized. Formally, we choose median matrices $med_{1j'} \dots med_{N_{max}j''}$, one per subject (for each subject i, med_{ij} could belong to any chosen strip s_{ij} containing ≥ 10 contiguous matrices), such that $\sum_{a=1}^{N_{max}-1} \sum_{b=a+1}^{N_{max}} CosineDistance(med_{aj'}, med_{bj''})$ is minimized. Here, $N_{max} =$ number of subjects with atleast one strip containing ≥ 10 contiguous matrices in cluster k. By considering the median (ordered by time) matrix, which corresponds to a window in time when the FC state was most stable, we ensure that we are closer to the true cluster center. By not weighting the medians with the length of the strips (number of matrices in the strip), we ensure that the objective function does not end up solely selecting small length strips.

3. We compute the center of the cluster k as the mean of the chosen median matrices. We then sort all the matrices (from all subjects) belonging to cluster k based on their cosine distances from the computed center.

4. We consider an appropriate percentile (chosen based on the first derivative of the cosine distance series - i.e. rate of change of cosine distance from the cluster center) of matrices from the above sorted order and use these windowed correlation matrices for centroid computation. The cluster centroids thus computed indicate quasi-stable FC states.

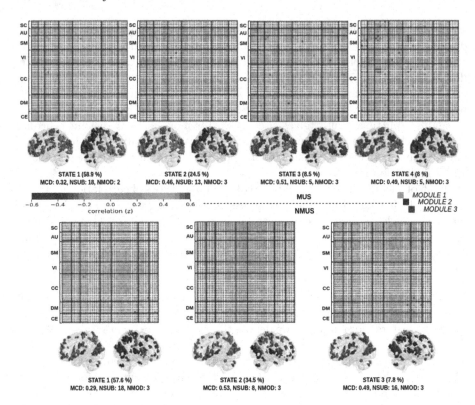

Fig. 3. Correlation heatmaps and Modularity partitions for the cluster centroids ordered by their percentage of occurrence. MCD values indicate the mean cosine distance of all the points in the cluster from the centroid. NSUB values indicate the number of subjects who have atleast one window in that cluster. NMOD indicates the number of modules in the Modularity partition.

The centroids of clusters which contain most of the data (states 1 and 2 for both Mus and NMus) are fully reproducible across bootstrap resamples of participants. To find community structures, the cluster centroids (ICN × ICN correlation map), which are indicative of recurring DFC states, were partitioned into modules using multiple runs of the Louvain modularity-maximization algorithm implemented in the Brain Connectivity Toolbox [19] (each module is thus a union of group level ICN spatial maps which were partitioned into that module).

3   Results and Discussion

Preliminary analysis (Fig. 3) reveals that overall, we observe more states in the musicians group, and a higher NSUB value (no. of subjects who have atleast one window in that cluster) on average for the non-musicians. This could be hypothesized to be attributed to greater similarities in listening strategies among

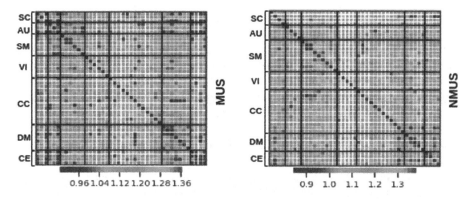

Fig. 4. Amplitude of FC oscillations - higher end of the scale indicates greater variability in connectivity over the course of time between the corresponding ICNs.

non-musicians (simpler sensory bottom-up listening) as against musicians who utilize finetuned top-down analytic listening strategies depending on their varied training methods, engendering differences in the underlying neural correlates.

In the most common state (state 1) of both the groups, for musicians we find a coupling between the visual and the sensorimotor regions (both are parts of module 2), while in non-musicians, the auditory regions and the sensorimotor regions lie in the same module (parts of module 2) with the visual regions lying in a separate module 3. This is in line with the action-perception coupling found in the musicians, wherein experience with a sensorimotor task such as instrument playing leads to a strong coupling of sensory (visual/auditory) and motor regions [17]. Studies have shown correlations in activations in sensorimotor regions and visual representation of music/instrument playing. On the other hand coupling between the auditory and sensorimotor regions in non-musicians can be hypothesized to be attributed to reactionary responses to acoustic features such as rhythm. For both groups, state 1 lacks coupling of the DMN regions, indicating that this state represents most of the active listening period with a high cognitive load.

For musicians, it can be suggested that state 3 is representative of the DMN related default state (DMN regions occur in the same module 3), representative of times of low cognitive load. State 2 and state 4 could be hypothesized to correspond to other states of active music listening, where state 4 is observable in a less number of subjects (training/prior experience dependent). State 2 presents greater integration of the auditory, visual and sensorimotor regions (module 1) as compared to state 1. State 4 indicates a few differences from state 1 - the subcortical regions along with the putamen/angular gyri and the inferior/superior frontal regions are grouped together (module 3).

For non-musicians, state 2 is primarily indicative of the grouping of DMN regions together (module 1), suggesting that this state corresponds to times of low cognitive load. The presence of this state for a reasonably large percentage of time (34% vs. 8%) as against in musicians, indicates a greater tendency to fall back to the default state in non-musicians in times of low auditory cognitive load. State 3 exhibits a separate module for visual regions (module 3), and also coupling of DMN regions. This, along with small-negative correlations of the visual regions with the other regions in the corresponding centroid FC matrix, calls for further investigation. It is to be noted that highlighting these differences in the temporal extent of DMN related activity (i.e. how long) is possible only due to the dynamic nature of the FC analysis adopted.

In terms of temporal variability (Fig. 4), musicians indicated a larger number of stable (less variable) (over the time course) ICN pairs, primarily associated with the Subcortical, Auditory, DMN and Cerebellar regions. Non-musicians exhibited a large number of fluctuating (more variable) connections between ICN pairs, primarily associated with the DMN, Visual and Sensorimotor regions. We note that pairwise ICN correlation variability timecourses have the potential to yield interesting information about the differences between the two groups.

To conclude, we are the first to analyze DFC in musicians and non-musicians in fMRI data collected in a naturalistic setting. We utilize spatial GICA, sliding time window correlations, and a deterministic agglomerative clustering of windowed correlation matrices to identify quasi-stable FC states in the two groups. We extend upon existing DFC analysis frameworks and present a method to choose appropriate matrices (corresponding to states which recur over time and across participants) for cluster centroid computation. Preliminary analysis indicate states with greater visuo-sensorimotor integration in musicians, larger presence of DMN states in non-musicians, and variability in states found in musicians due to differences in training and prior experiences. Further analysis to unearth more details about these states is called for.

Acknowledgements. This work was supported by the Academy of Finland (project numbers 272250 and 274037) and the Danish National Research Foundation (DNRF117).

References

1. Allen, E.A., Damaraju, E., Plis, S.M., Erhardt, E.B., Eichele, T., Calhoun, V.D.: Tracking whole-brain connectivity dynamics in the resting state. Cereb. Cortex **24**(3), 663–676 (2013)
2. Alluri, V., Toiviainen, P., Burunat, I., Kliuchko, M., Vuust, P., Brattico, E.: Connectivity patterns during music listening: evidence for action-based processing in musicians. Hum. Brain Mapp. **38**(6), 2955–2970 (2017)
3. Alluri, V., Toiviainen, P., Jääskeläinen, I.P., Glerean, E., Sams, M., Brattico, E.: Large-scale brain networks emerge from dynamic processing of musical timbre, key and rhythm. NeuroImage **59**, 3677–3689 (2012)
4. Angulo-Perkins, A., Aubé, W., Peretz, I., Barrios, F.A., Concha, L.C.B.: Music listening engages specific cortical regions within the temporal lobes: differences between musicians and non-musicians. Cortex **59**, 126–137 (2014)
5. Blondel, V.D., Guillaume, J.L., Lambiotte, R., Lefebvre, E.: Fast unfolding of communities in large networks. J. Stat. Mech.: Theory Exp. **2008**(10), P10008 (2008)

6. Braun, U., et al.: Dynamic reconfiguration of frontal brain networks during executive cognition in humans. Proc. Natl. Acad. Sci. U. S. A. **112**, 11678–11683 (2015)
7. Burunat, I., Tsatsishvili, V., Brattico, E., Toiviainen, P.: Coupling of action-perception brain networks during musical pulse processing: evidence from region-of-interest-based independent component analysis. Front. Hum. Neurosci. **11**, 230 (2017)
8. Calhoun, V.D., Adali, T., Pearlson, G.D., Pekar, J.J.: A method for making group inferences from functional MRI data using independent component analysis. Hum. Brain Mapp. **14**(3), 140–151 (2001)
9. Damaraju, E., et al.: Dynamic functional connectivity analysis reveals transient states of dysconnectivity in schizophrenia. NeuroImage: Clin. **5**, 298–308 (2014)
10. Fauvel, B., et al.: Morphological brain plasticity induced by musical expertise is accompanied by modulation of functional connectivity at rest. NeuroImage **90**, 179–188 (2014)
11. Gaser, C., Schlaug, G.: Brain structures differ between musicians and non-musicians. J. Neurosci. **23**(27), 9240–9245 (2003)
12. Gonzalez-Castillo, J., Bandettini, P.A.: Task-based dynamic functional connectivity: recent findings and open questions. NeuroImage **180**, 526–533 (2018). Brain Connectivity Dynamics
13. Hutchinson, S., Lee, L.H.L., Gaab, N., Schlaug, G.: Cerebellar volume of musicians. Cereb. Cortex **13**(9), 943–949 (2003)
14. Imfeld, A., Oechslin, M., Meyer, M., Loenneker, T., Jäncke, L.: White matter plasticity in the corticospinal tract of musicians: a diffusion tensor imaging study. NeuroImage **46**, 600–607 (2009)
15. Koelsch, S., et al.: The roles of superficial amygdala and auditory cortex in music-evoked fear and joy. NeuroImage **81**, 49–60 (2013)
16. Maneshi, M., Vahdat, S., Gotman, J., Grova, C.: Validation of shared and specific independent component analysis (SSICA) for between-group comparisons in fMRI. Front. Neurosci. **10**, 417 (2016)
17. Novembre, G., Keller, P.E.: A conceptual review on action-perception coupling in the musicians' brain: what is it good for? Front. Hum. Neurosci. **8**, 603 (2014)
18. Preti, M.G., Bolton, T.A.W., Ville, D.V.D.: The dynamic functional connectome: state-of-the-art and perspectives. NeuroImage **160**, 41–54 (2017)
19. Rubinov, M., Sporns, O.: Complex network measures of brain connectivity: uses and interpretations. NeuroImage **52**(3), 1059–1069 (2010). Computational Models of the Brain
20. Toiviainen, P., Alluri, V., Brattico, E., Wallentin, M., Vuust, P.: Capturing the musical brain with lasso: dynamic decoding of musical features from fMRI data. NeuroImage **88**, 170–180 (2014)
21. Wilkins, R.W., Hodges, D.A., Laurienti, P.J., Steen, M.L., Burdette, J.H.: Network science and the effects of music preference on functional brain connectivity: from Beethoven to Eminem. Sci. Rep. **4**, 6130 (2014)
22. Zimek, A., Schubert, E., Kriegel, H.P.: A survey on unsupervised outlier detection in high-dimensional numerical data. Stat. Anal. Data Min.: ASA Data Sci. J. **5**(5), 363–387 (2012)

Brain Big Data Analytics, Curation and Management

Imaging EEG Extended Sources Based on Variation Sparsity with L_1-norm Residual

Furong Xu, Ke Liu$^{(\boxtimes)}$, Xin Deng, and Guoyin Wang

Chongqing Key Laboratory of Computational Intelligence,
Chongqing University of Posts and Telecommunications, Chongqing, China
liuke@cqupt.edu.cn

Abstract. Reconstructing the locations and extents of cortical activities accurately from EEG signals plays an important role in neuroscience research and clinical surgeries. However, due to the ill-posed nature of EEG source imaging problem, there exist no unique solutions for the measured EEG signals. Additionally, evoked EEG is inevitably contaminated by strong background activities and outliers caused by ocular or head movements during recordings. To handle these outliers and reconstruct extended brain sources, in this paper, we have proposed a robust EEG source imaging method, namely L_1-norm Residual Variation Sparse Source Imaging (L_1R-VSSI). L_1R-VSSI employs the L_1-loss for the residual error to alleviate the outliers. Additionally, the L_1-norm constraint in the variation domain of sources is implemented to achieve globally sparse and locally smooth solutions. The solutions of L_1R-VSSI is obtained by the alternating direction method of multipliers (ADMM) algorithm. Results on both simulated and experimental EEG data sets show that L_1R-VSSI can effectively alleviate the influences from the outliers during the recording procedure. L_1R-VSSI also achieves better performance than traditional L_2-norm based methods (e.g., wMNE, LORETA) and sparse constraint methods in the original domain (e.g., SBL) and in the variation domain (e.g., VB-SCCD).

Keywords: EEG inverse problem · Variation sparseness · Outliers · ADMM

1 Introduction

Electroencephalogram (EEG) are widely used techniques in cognitive neuroscience and clinical surgeries due to their non-invasiveness and high temporal

Supported in part by the National Natural Science Foundation of China under Grants 61703065, 61876027. Chongqing Research Program of Application Foundation and Advanced Technology under Grant cstc2018jcyjAX0151, the Science and Technology Research Program of Chongqing Municipal Education Commission under Grant KJQN201800612 and KJQN201800638.

P. Liang et al. (Eds.): BI 2019, LNAI 11976, pp. 95–104, 2019.
https://doi.org/10.1007/978-3-030-37078-7_10

resolution. The task to reconstruct the corresponding cortical neural activities from EEG signals on the scalp surface is termed as EEG source imaging. Using EEG source imaging, we can estimate the extents and locations of brain active regions, which provide important auxiliary information for neuroscience research and medical treatments.

To reconstruct brain activities, the distributed source model divides the cerebral cortex into several triangular grids, each of which represents a dipole source [8]. The amplitude of each diploe source can be calculated by solving a linear inverse problem. However, the number of the candidate dipole sources largely outnumbered that of EEG sensors. Hence, the linear inverse problem is highly ill-posed, i.e., there are infinite source solutions to satisfy EEG signals. To obtain a unique solution, appropriate regularization operator is needed to narrow the solution space [8,9,16].

Typical EEG source imaging methods employ the L_2-norm regularizer, such as the weighted minimum norm estimate (wMNE) [4] and the low-resolution electromagnetic tomography (LORETA) [11]. However, the estimations of L_2-norm based methods are overly diffused and spread widely [8]. To improve the spatial resolution, sparse constrained methods based on L_p-norm ($0 < p \leq 1$) regularizer [1] and sparse Bayesian learning (SBL) [15] framework are developed. However, the solutions of conventional sparse constrained methods are too sparse and provide little information about source extents. To estimate the locations and extents of brain activities, several studies have proposed to implement sparse constraint on the transformed domains, such as the variation, wavelet and Laplacian domains [3,5,10,16], which can provide more information about the source extents than the sparse constraints in the original source domains.

Most of the aforementioned methods assume that the measurement noise satisfies Gaussian distribution and employ the L_2-norm to measure the residual error. Nevertheless, the EEG signals are inevitably contaminated by the strong background activities and outliers caused by ocular or head movements during recording procedure. The L_2-norm may not be suitable to represent these outliers. To handle these outliers in the EEG signals and estimate the extended cortical activities, as in [1], we propose a robust EEG source imaging method, namely L_1-norm Residual Variation Sparse Source Imaging (L_1R-VSSI). L_1R-VSSI employs the L_1-norm to measure the residual error. In addition, to estimate the locations and extents of sources, L_1R-VSSI attains sparseness of sources on the variation domain by incorporating the L_1-norm constraint. The cortical sources of L_1R-VSSI are efficiently obtained by the alternating direction method of multipliers (ADMM) algorithm [2].

The remaining of the paper is organized as follows. In Sect. 2, we present details of the proposed EEG source imaging algorithm. In Sect. 3, the simulation design and evaluation metrics are presented. Section 4 presents the results on simulated and real EEG data, followed by a brief conclusion in Sect. 5.

2   Methods

2.1   Background

The relationship between the EEG recordings and cortical sources can be expressed as [8]

$$b = Ls + \varepsilon \tag{1}$$

where $b \in \mathbb{R}^{d_b \times 1}$ is EEG recording, $s \in \mathbb{R}^{d_s \times 1}$ is the unknown source vector. $L \in \mathbb{R}^{d_b \times d_s}$ is the lead-field matrix that describes how a unit source of a certain candidate location is related to the EEG measurements. d_b and d_s denote the number of sensors and candidate sources respectively. ε denotes the measurement noise vector.

To estimate the extents and locations of sources, as in [5,10], we assume that the sources are sparse in the variation domain. Each element in the variation domain is the difference of amplitudes between two adjacent dipole sources. When the amplitudes of sources are uniform, the non-zero vales of the variation sources are largely expected to occur on the boundaries between the active and inactive areas. Hence, we can estimate the sparseness on the variation domain to obtain the extents and locations of cortical activities. To obtain the variation sources, a variation operator $V \in \mathbb{R}^{P \times d_s}$ is defined as [5,10]

$$V = \begin{bmatrix} v_{11} & v_{12} & \cdots & v_{1d_s} \\ v_{21} & v_{22} & \cdots & v_{2d_s} \\ \vdots & \vdots & \ddots & \vdots \\ v_{P1} & v_{P2} & \cdots & v_{Pd_s} \end{bmatrix} \tag{2}$$

$$\begin{cases} v_{pi} = 1, v_{pj} = -1, i < j; \text{if source } i, j \text{ share edge } p \\ v_{pi} = 0; \quad \text{otherwise} \end{cases}$$

where P denotes the number of edges of triangular grids, which are defined by the source model. The pth row of V corresponds to edge p. For each row of V, only two elements corresponding to the two dipoles that share edge p are non-zero (i.e., 1, -1). The pth row of the variation sources $u = Vs \in \mathbb{R}^{P \times 1}$, u_p, indicates the difference of dipoles that share the pth edge.

2.2 L_1R-VSSI: L_1-norm Residual Variation Sparse Source Imaging

Usually, the EEG measurement noise ε is assumed to satisfy Gaussian distribution. And the fitting error is measured by the L_2-norm for most EEG source imaging methods. However, the EEG signals are inevitably influenced by the artifacts induced by the head movement or eye blinks. The L_2-norm for the fitting error will exaggerates the effect of these outliers. To handle these outliers or artifacts in EEG, our proposed EEG source imaging method, L_1R-VSSI, employs the L_1-norm to measure the fitting error. Combining the variation sparse constraint, the solution of L_1R-VSSI is obtained as

$$\hat{s} = \arg\min_s \|Ls - b\|_1 + \lambda \|Vs\|_1 \tag{3}$$

where $\lambda > 0$ is the regularization parameter. In this work, the ADMM algorithm is employed to solve formulation (3).

We first rewrite Eq. (3) as

$$\hat{s} = \arg\min_{s} \|e\|_1 + \lambda\|u\|_1$$

$$s.t., \quad e = Ls - b, u = Vs \qquad (4)$$

Then the augmented Lagrangian function of the constrained optimization problem (4) is

$$\mathcal{L}(s,e,u,y,z) = \|e\|_1 + \lambda\|u\|_1 + y^\top(Ls - b - e) + \frac{\rho_1}{2}\|Ls - b - e\|_2^2$$
$$+ z^\top(Vs - u) + \frac{\rho_2}{2}\|Vs - u\|_2^2 \qquad (5)$$

where $\rho_1 > 0$ and $\rho_2 > 0$ are penalty parameters, $y \in \mathbb{R}^{d_b \times 1}$ and $z \in \mathbb{R}^{P \times 1}$ are the Lagrangian multipliers. Since $\mathcal{L}$ is separable with respect the variables (s,e,u), we can obtain (s,e,u) with three subproblems.

$$\hat{s}^{(k+1)} = \arg\min_{s} \mathcal{L}\left(s, e^{(k)}, u^{(k)}, y^{(k)}, z^{(k)}\right)$$
$$= \left(\rho_1 L^\top L + \rho_2 V^\top V\right)^{-1}\left[\rho_1 L^\top\left(b + e^{(k)}\right)\right.$$
$$\left. + \rho_2 V^\top u^{(k)} - L^\top y^{(k)} - V^\top z^{(k)}\right] \qquad (6)$$

$$\hat{e}^{(k+1)} = \arg\min_{e} \mathcal{L}\left(s^{(k+1)}, e, u^{(k)}, y^{(k)}, z^{(k)}\right)$$
$$= S_{\frac{1}{\rho_1}}\left(Ls^{(k+1)} - b + \frac{1}{\rho_1}y^{(k)}\right) \qquad (7)$$

$$\hat{u}^{(k+1)} = \arg\min_{u} \mathcal{L}\left(s^{(k+1)}, e^{(k+1)}, u, y^{(k)}, z^{(k)}\right)$$
$$= S_{\frac{\lambda}{\rho_2}}\left(Vs^{(k+1)} + \frac{1}{\rho_2}z^{(k)}\right) \qquad (8)$$

where $S_\kappa(a)$ is

$$S_\kappa(a) = \begin{cases} a - \kappa, & a > \kappa \\ 0, & |a| < \kappa \\ a + \kappa, & a < -\kappa \end{cases} \qquad (9)$$

The Lagrangian multipliers y, z are updated using the dual ascent method [2]

$$\hat{y}^{(k+1)} = y^{(k)} + \rho_1\left(Ls^{(k+1)} - b - e^{(k+1)}\right) \qquad (10)$$
$$\hat{z}^{(k+1)} = z^{(k)} + \rho_2\left(Vs^{(k+1)} - u^{(k+1)}\right) \qquad (11)$$

where $x^{(k)}$ is the value of x at kth iteration.

In summary, to obtain the source estimations, L_1R-VSSI alternatively updates the variables (s,e,u) and the Lagrangian multipliers (y,z), until the relative change of s reaches a user-specified tolerance (e.g., 10^{-4}). Through our simulations, e, u, y, z are all initialized to 0. The penalty parameters ρ_1, ρ_2 are determined by cross-validation [14]. For the simulation experiment of $d_s = 6002$, $d_b = 62$, $P = 17977$, it reaches convergence after approximately 200 iterations, which takes about 2 min.

3   Simulation Design and Evaluation Metrics

To test the performance of L_1R-VSSI, a series of Monte-Carlo numerical simulations were set up to compare L_1R-VSSI and four imaging methods: two widely used L_2-norm based methods, i.e., wMNE [4] and LORETA [11], and sparse constraint methods in the original domain (SBL [15]) and in the variation domain (variation-based sparse cortical current density (VB-SCCD) [5]). Additionally, L_1R-VSSI was also applied to one real clinical EEG data.

3.1   Simulation Design

In the numerical simulations, a three-shell head model is obtained using Brainstorm [13]. There are 6002 triangle grids evenly distributed on the cortical surface, each of which is a dipole source. The dipole orientation is fixed to be normal to the cortical surface. The lead-field matrix L was constructed using Brainstorm with the sensor configuration of the 64-channel Neuroscan Quik-cap system (two channels are not EEG electrodes, hence, $L \in \mathbb{R}^{62 \times 6002}$).

A triangle grid is randomly selected as a seed point. The adjacent triangles are gradually added to form an extended source. After multiplying the source vector with the lead-field matrix, the cortical activities are transformed into EEG electrode recordings. To simulate the actual EEG signals, both Gaussian noise and outliers are added to the EEG electrode recordings. The noise level is described by signal-to-noise ratio (SNR), which is defined as SNR $= 10 \log_{10} \left[\frac{\sigma^2(Ls)}{\sigma^2(\varepsilon)} \right]$. where Ls is raw EEG data without noise and ε is mixture measurement noise (Gaussian and outliers). $\sigma^2(x)$ is the variance of x.

As in [1], the mixture measurement noise is defined as

$$\varepsilon = \frac{\varepsilon_1}{\sigma(\varepsilon_1)} \left[10^{(-\frac{SNR}{20})} \right] \sigma(Ls) \tag{12}$$

where ε_1 denote the outliers and σ is the standard deviation.

To generate the mixture noise, we first generate the outliers $\varepsilon_1 \in \mathbb{R}^{d_b \times 1}$, where each element of ε_1 obeys Gaussian distribution $\mathcal{N}(\mu + 10\sigma^2, \sigma^2)$ [1]. μ and σ^2 are the mean and variance of the true EEG Ls across all channels at the single time slice, respectively. Then, given a specified SNR, the mixture noise is generated by Eq. (12) and added to the true EEG recordings Ls.

In the numerical Monte-Carlo experiments, the performance of the proposed method are verified in two cases: (1) One extended source with different extents (i.e., 0.8, 4, 8, 12, 18 cm^2); (2) EEG signals with different SNRs (i.e., 0, 5, 10, 15 dB). For each case, 100 simulation experiments were conducted to ensure that the simulated extended sources could cover most areas of the brain.

3.2   Evaluation Metrics

The performance of imaging algorithms are evaluated by four evaluation metrics. (1) The area under the receiver operating characteristic (ROC) curve (AUC), which assesses the sensitivity and specificity of imaging algorithms [7]; (2) Spatial dispersion (SD), which depicts that the level of the spatial dispersion of reconstructed source [3]; (3) Distance of localization error (DLE), which describes the location error of the estimated source activities [16]; (4) Relative mean square error (RMSE), which estimates the amplitudes accuracy of the reconstructed sources [9]. Larger AUC, smaller SD, DLE and RMSE values indicate that the method has better performance. The detailed computation of these metrics can refer to the supplementary document in [10]. To image the estimated sources, the absolute value of the recovered sources at specified time points are shown, thresholded by the Otsu's method [7, 10].

4   Results

4.1   Results for Various Extents

Figure 1 shows performance metrics under different extents. As the source extents increases, the AUC values of wMNE, SBL, VB-SCCD, L_1R-VSSI gradually decrease. In contrast, LORETA's AUC value increases, indicating the potential of LORETA to estimate large extents' sources. The SD, DLE, RMSE values of wMNE, LORETA, VB-SCCD and L_1R-VSSI gradually reduced, but L_1R-VSSI get the smallest values, which means that L_1R-VSSI have more accurate reconstructed solutions, smaller spatial dispersion and location errors. Due to the sparse constraint of SBL, although it gets the smallest SD, DLE values, its overly focal results also lead to a gradually increasing RMSE value. Comprehensively comparing the results of four evaluation metrics, the L_1R-VSSI method shows better performance than the other four methods under different source extents.

Figure 2 depicts the imaging maps of reconstructed sources under different extents. The first column in Fig. 2 shows the simulated extended sources, and the remaining columns show the reconstructed sources estimated by the five imaging methods respectively. From the Fig. 2, we can see that SBL can reconstruct the focal sources perfectly. However, for large extents' sources, it can only accurately locate the center of the sources, but the estimated extents are overly focused. In contrast, the solutions of wMNE and LORETA are too blurred compared to the simulated sources. Because the addition of variation operator, the blurriness of the imaging results of VB-SCCD solution is greatly reduced and more sensitive to the extents of the simulated sources. Nevertheless, compared to VB-SCCD, L_1R-VSSI approach's recovered solutions have more precise description for extents and strengths. This is because the use of L_1-loss of the residual error reduces the influence of artificial noise. Combining the results of Figs. 1 and 2, L_1R-VSSI performs better than wMNE, LORETA, SBL and VB-SCCD.

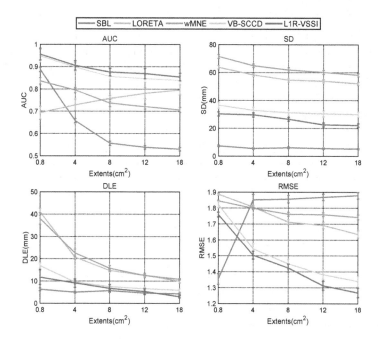

Fig. 1. Evaluation metrics of different source extents. The data is the results of 100 Monte Carlo simulations and is described as Mean ± SEM (SEM: standard error of mean). The SNR is 5 dB.

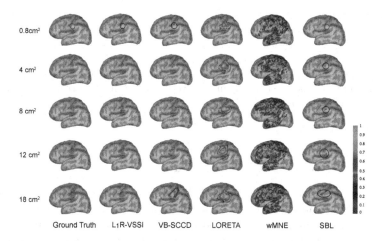

Fig. 2. Imaging results for different extents of extended sources. Source activity maps show the absolute value of the extended sources' activities. The threshold is determined using Otsu's method. The SNR is 5 dB. Some sources are circled for illustration purposes.

4.2   Results for Different SNRs

Figure 3 shows the performance evaluation metrics under different SNRs (0, 5, 10, 15 dB). All algorithms are significantly affected by the noise level. As the SNR increases, these algorithms' AUC value gradually increases, and the RMSE, SD, and DLE value decrease. Under comprehensive comparison, L_1R-VSSI has a larger AUC value, smaller SD, RMSE, DLE value, so the performance is better than other approaches.

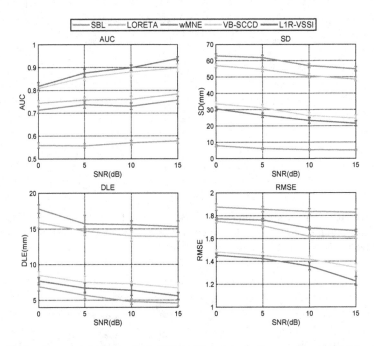

Fig. 3. Evaluation metrics of various SNRs. The data is the results of 100 Monte Carlo simulations and is described as Mean $\pm$ SEM (SEM: standard error of mean). The extents of source is about 8 cm^2.

4.3   Application of Epilepsy EEG Data

To show the practical performance of L_1R-VSSI, we further apply the proposed method to the Brainstorm public epilepsy data[1]. The average spikes across 58 trials is used to estimate sources, which are presented in Fig. 4(a). We calculate the lead-field matrix based head model obtained from the head anatomical structure of the subject. The EEG data was collected from a patient who suffered from focal epilepsy with focal sensory, dyscognitive and secondarily generalized seizures since the age of eight years. After the resection of the epileptogenic area,

---

[1] https://neuroimage.usc.edu/brainstorm/DatasetEpilepsy.

postoperative investigation showed the patient no epilepsy in 5 years. In [6], it has located that the patient's epileptogenic area is left frontal, which was estimated by invasive EEG. Figure 4(b) shows the imaging results of various algorithms at the peak time point (i.e., 0 s). Comparing the results in Fig. 4(b) with the analysis in [6], we can find that these algorithms all locate the epileptogenic area. However, the epileptogenic area identified by LORETA and wMNE are overly diffused, even covering many other areas of the brain scalp. Conversely, SBL can only obtain some point sources in the epileptogenic area. For the variation sparse constraint based methods, the estimations by L_1R-VSSI and VB-SCCD is well accordant with previous reports [6,12]. However, the solution of L_1R-VSSI shows more clear boundaries than that of VB-SCCD.

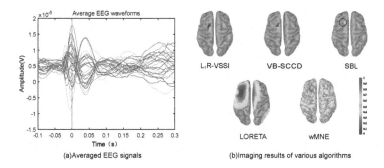

Fig. 4. Application of the public epilepsy data. (a) is averaged EEG signals across 58 trials. (b) is the imaging maps, showing the absolute value of the source activities at the peak time point. The imaging threshold was determined by Otsu's method. SBL's estimation is circled for illustration purposes.

5   Conclusions

We have proposed a robust EEG source estimation algorithm, L_1R-VSSI, which reconstructs extended sources using L_1-loss of the residual error and sparse constraints in the variation domains. The solution of L_1R-VSSI is efficiently obtained by ADMM. Numerical results indicate that L_1R-VSSI can effectively alleviate the influence of outliers and achieves better performance (larger AUC value and smaller SD, DLE, RMSE value) than wMNE, LORETA, SBL and VB-SCCD in terms of reconstructing extended sources. However, due to page limit, for the experimental data analysis, we only applied L_1R-VSSI to analyze the epilepsy EEG data from Brainstorm. In our future work, we will also apply the proposed method to more clinical data to verify the performance, which will promote the development of neuroscience research and clinical medicine. Additionally, since the minimization of sources in the variation domain does not limit the global energy, L_1R-VSSI tends to underestimate the amplitude of sources [10,16]. Our future work will also consider adding regularizers in other transform domains to improve the accuracy of amplitude.

References

1. Bore, J.C., et al.: Sparse EEG source localization using LAPPS: least absolute l-P ($0 < p < 1$) penalized solution. IEEE Trans. Biomed. Eng. **66**, 1927–1939 (2018)
2. Boyd, S., Parikh, N., Chu, E., Peleato, B., Eckstein, J., et al.: Distributed optimization and statistical learning via the alternating direction method of multipliers. Found. Trends® Mach. Learn. **3**(1), 1–122 (2011)
3. Chang, W.T., Nummenmaa, A., Hsieh, J.C., Lin, F.H.: Spatially sparse source cluster modeling by compressive neuromagnetic tomography. Neuroimage **53**(1), 146–160 (2010)
4. Dale, A.M., Sereno, M.I.: Improved localizadon of cortical activity by combining EEG and MEG with MRI cortical surface reconstruction: a linear approach. J. Cognit. Neurosci. **5**(2), 162–176 (1993)
5. Ding, L.: Reconstructing cortical current density by exploring sparseness in the transform domain. Phys. Med. Biol. **54**(9), 2683–2697 (2009)
6. Dümpelmann, M., Ball, T., Schulze-Bonhage, A.: sLORETA allows reliable distributed source reconstruction based on subdural strip and grid recordings. Hum. Brain Mapp. **33**(5), 1172–1188 (2012)
7. Grova, C., Daunizeau, J., Lina, J.M., Bénar, C.G., Benali, H., Gotman, J.: Evaluation of EEG localization methods using realistic simulations of interictal spikes. Neuroimage **29**(3), 734–753 (2006)
8. He, B., Sohrabpour, A., Brown, E., Liu, Z.: Electrophysiological source imaging: a noninvasive window to brain dynamics. Annu. Rev. Biomed. Eng. **20**, 171–196 (2018)
9. Liu, K., Yu, Z.L., Wu, W., Gu, Z., Li, Y., Nagarajan, S.: Bayesian electromagnetic spatio-temporal imaging of extended sources with Markov random field and temporal basis expansion. Neuroimage **139**, 385–404 (2016)
10. Liu, K., Yu, Z.L., Wu, W., Gu, Z., Li, Y., Nagarajan, S.: Variation sparse source imaging based on conditional mean for electromagnetic extended sources. Neurocomputing **313**, 96–110 (2018)
11. Pascual-Marqui, R.D., Michel, C.M., Lehmann, D.: Low resolution electromagnetic tomography: a new method for localizing electrical activity in the brain. Int. J. Psychophysiol. **18**(1), 49–65 (1994)
12. Sohrabpour, A., Lu, Y., Worrell, G., He, B.: Imaging brain source extent from EEG/MEG by means of an iteratively reweighted edge sparsity minimization (IRES) strategy. Neuroimage **142**, 27–42 (2016)
13. Tadel, F., Baillet, S., Mosher, J.C., Pantazis, D., Leahy, R.M.: Brainstorm: a user-friendly application for MEG/EEG analysis. Comput. Intell. Neurosci. **2011**, 8:1–8:13 (2011)
14. Wen, F., Pei, L., Yang, Y., Yu, W., Liu, P.: Efficient and robust recovery of sparse signal and image using generalized nonconvex regularization. IEEE Trans. Comput. Imaging **3**(4), 566–579 (2017)
15. Wipf, D., Nagarajan, S.: A unified Bayesian framework for MEG/EEG source imaging. Neuroimage **44**(3), 947–966 (2009)
16. Zhu, M., Zhang, W., Dickens, D.L., Ding, L.: Reconstructing spatially extended brain sources via enforcing multiple transform sparseness. Neuroimage **86**, 280–293 (2014)

Image-Assisted Discrimination Method for Neurodevelopmental Disorders in Infants Based on Multi-feature Fusion and Ensemble Learning

Xiaohui Dai[1], Shigang Wang[1(✉)], Honghua Li[2], Haichun Yue[1], and Jiayuan Min[1]

[1] College of Communication Engineering, Jilin University,
Changchun 130012, China
wangshigang@vip.sina.com
[2] Department of Developmental and Behavioral Pediatrics,
The First Hospital of Jilin University, Changchun 130021, China

Abstract. Premature infants have a significantly increased risk of developing severe neurodevelopmental disorders such as cerebral palsy and mental retardation due to some congenital defects at birth. During early infancy, distinct motion patterns occur which are highly predictive for later disability. The clinical observations of these forms of exercise can be record as parameters. In this paper, we used Kernel Correlation Filter (KCF) to track the trajectories of an infant's limbs. Then, the obtained trajectories are analyzed in the wavelet domain and power spectrum domain, and integrated the features into the Ensemble Learning classification, the classification results are weighted and comprehensively judged to determine whether the infant's neurodevelopment is normal and whether early intervention is needed.

Keywords: Neurodevelopmental disorder · Ensemble Learning · Target-tracking · Multi-feature fusion · Stacking

1 Introduction

The morphology and physiological functions of premature infants are immature, causing them to have low immunity, and making it easy for them to suffer from infection, intracranial hemorrhage, hypoxic ischemic encephalopathy, hyperbilirubinemia, and encephalopathy, which can cause non-progressive brain injury and is one of the most common causes of cerebral palsy. At the same time, there are other types of neurodevelopmental disorders associated with premature birth, such as mental retardation, visual and auditory disorders, and mild neurological dysfunction. Because the development of brain tissue in premature infants is not mature, the cerebral cortex is thin, and the myelin sheath is not completely formed and has strong plasticity. Thus, any motor developmental disorder caused by brain injury is in the primary stage, and any abnormal posture and muscle tension have not been immobilized. Consequently, early rehabilitation treatment can provide benign stimulation, and the brain potential

© Springer Nature Switzerland AG 2019
P. Liang et al. (Eds.): BI 2019, LNAI 11976, pp. 105–114, 2019.
https://doi.org/10.1007/978-3-030-37078-7_11

can be fully developed, allowing the damaged brain to mature and differentiate, which can reduce the degree of brain damage. Therefore, early intervention in small infants with abnormal motor patterns is essential to improve their neurodevelopmental outcomes.

The Austrian developmental neuroscientist Professor Prechtl [1] first proposed in 1993 that there are spontaneous whole body movements called general movements (GMs) in the early stage of human development. It is the most frequently occurring and most complex spontaneous exercise pattern. It first appears in the 9th week of pregnancy, and will last for 5 to 6 months after birth. It is a very effective means of assessing the function of the young nervous system. In clinical practice, the evaluation of the whole body exercise quality of infants often requires a doctor with relevant evaluation qualifications and relatively rich experience. In this regard, we examine whether computer vision can be used instead of human evaluation to improve the efficiency and accuracy.

For example, a method was proposed for making an infant behavioral judgment by calculating the infant's speed, acceleration, and centroid parameters during exercise [2]. Meinecke et al. [3] used seven cameras to analyze the movement of a baby by capturing three-dimensional markers attached to the newborn.

Without the aid of sensors and other devices that need to be attached to the infant's body, we propose a method for classifying the infant's limb movement behavior based on the infant's exercise video to determine whether the infant's behavior is abnormal. Our aim is to track the limbs of the baby by KCF. We analyze the obtained motion trajectory and use differential samples for classification training to determine whether the infant behavior is abnormal. Wavelet analysis and wavelet power spectrum analysis are performed on the obtained motion trajectory. In other words, information about the trajectories of the infant's limbs is obtained in both the wavelet and power spectrum domains. The wavelet and wavelet power spectrum domain information is divided into samples, and the stacking method in Ensemble Learning is used for classification training. The weighted fusion algorithm is used to distribute the weights to obtain the highest accuracy.

2 Methods

2.1 Motion Positioning and Tracking

We used Kernel Correlation Filter (KCF) [5] to track the trajectories of an infant's limbs. KCF uses the given samples to train a discriminant classifier to determine whether the target or surrounding background information is tracked. The samples are mainly collected using a rotation matrix, and the algorithm is accelerated using Fast Fourier Transform. Figure 1 shows the target's limbs and overall body trajectory.

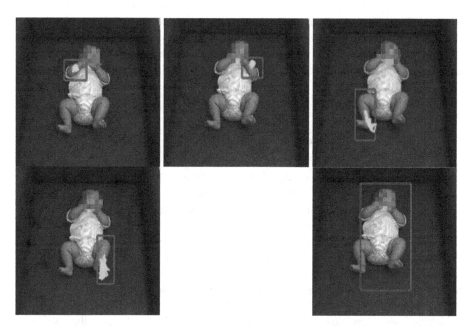

Fig. 1. Infant target tracking, where red frame shows target position, and green line indicates motion track. (Color figure online)

2.2 Infant Trajectory Analysis

After the target tracking of the infant's limbs and whole body, we obtain the motion trajectory under the X and Y axes in Fig. 2. Because normal infants and abnormal infants showed little difference in their motion trajectories under the X axis [11], the following studies are aimed at the trajectory under the Y axis. For the information contained in the infant's trajectory, we divide the wavelet domain and power spectrum into two parts for information extraction.

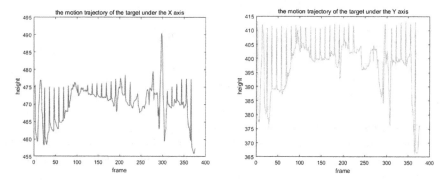

Fig. 2. Motion trajectory under coordinate axes. The left picture shows the motion trajectory of the target under the X axis, and the right picture shows the motion trajectory of the target under the Y axis.

Wavelet Transform Analysis of Trajectory Information

Because the obtained Y axis trajectory waveform is a non-stationary signal, in order to extract further information from the waveform, we consider using the discrete wavelet transform, which has both frequency information and time information, and the wavelet base can be flexibly selected.

The formula of the wavelet transform is as Eq. 1:

$$WT(a, \tau) = \frac{1}{\sqrt{a}} \int_{-\infty}^{\infty} f(t) * \varphi\left(\frac{t - \tau}{a}\right) dt \tag{1}$$

where f(t) represents the original signal. The wavelet transform has two variables: scale a and translation τ. Scale a controls the expansion and contraction of the wavelet function, and the translation amount, τ, controls the translation of the wavelet function.

One advantage of wavelet transform is that it has multi-resolution characteristics (also called multi-scale). It can use coarse to fine signal steps, and in a place where the waveform is relatively flat, it can be distinguished using a large scale. For areas where the waveform changes are severe, it can use small scales to resolve these, without ignoring the details of the waveform information as much as possible.

The waveform has high-frequency and low-frequency parts. Thus, we filter the waveform using low-pass and high-pass filters (Eqs. 2 and 3), and then different waveform transforms can be obtained for different scale spaces:

$$A_{j+1}(n) = H(n) * A_j(n) \tag{2}$$

And

$$A_{j+1}(n) = \sum_k H(k)A_j(2n - k)$$
$$D_{j+1}(n) = G(n) * A_j(n)$$

And

$$D_{j+1}(n) = \sum_k G(k)A_j(2n - k) \tag{3}$$

where H(n) and G(n) are the sequences of the tap coefficients of the low-pass and high-pass filters corresponding to the wavelet function, respectively, $A_{j+1}(n)$ and $D_{j+1}(n)$ represent the original approximate signal and detailed signal of the waveform, respectively. Figure 3 shows the waveform approximation of a certain extracted waveform.

Motion Trajectory Power Information Analysis

According to the power spectrum information, we can see how much energy the signal contains in the time frequency unit. The power spectrum information of the signal can be obtained by calculating the square of the amplitude spectrum on the original signal (Eq. 4). Figure 4 shows the power spectrum obtained for a certain waveform.

$$p = \left| f(n)^2 \right| \tag{4}$$

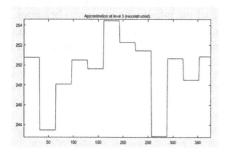

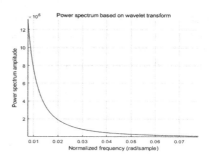

Fig. 3. Waveform approximation at level 5.   **Fig. 4.** Power spectrum.

2.3 Stacking Classification

Stacking is a method of Ensemble Learning, it used the output of a series of models (also called the base model) as the input of other models. This method implements the cascading of the model, that is, the output of the first layer is used as the input of the second layer model, and the output of the second layer is used as the input of the third layer model, and so on, and the output of the final layer model is the final result. In this paper, we used two layers of stacking.

3 Experiments

3.1 Selection of Test Subjects

The infant videos used in our experiments were taken with a digital video camera. During recording, the subject was placed in a supine position on a warm box or mat. The infant's limbs remained exposed and active, without crying, irritability, continued snoring, and the use of a pacifier avoided during recording. We selected a total of 120 samples, of which 60 were normal-behavior infants and 60 were abnormal babies, with ages between 10 and 20 weeks.

3.2 Motion Track

Using the MATLAB software, the infant video was first pre-processed, including uniform naming and segmentations with the same duration, to ensure that the subsequent feature dimensions were the same. Then, the infant in each video was targeted, and the tracking trajectory results were saved.

3.3 Motion Trajectory Analysis

The following experiment was performed for a motion trajectory below the Y axis. We can obtain the wavelet approximation using by wavelet transform, as shown in Fig. 5, which corresponds to the wavelet waveform approximation of the infant's limbs, where green indicates a sample of 30 clinical GMs used to evaluate abnormal infants and red indicates a sample of 30 clinical GMs used to evaluate normal infants. It can be seen that the wavelet waveform approximation map of the infant with abnormal behavior is more concentrated within a certain range than the wavelet waveform approximation of the infant with normal behavior, while the motion waveform of the normal infant is more scattered and irregular.

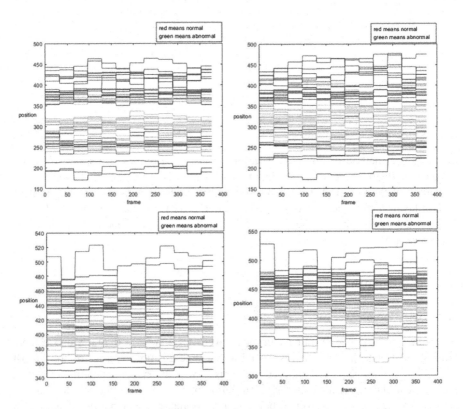

Fig. 5. Wavelet waveform approximation of infant's limbs, where green indicates sample of 30 clinical GMs used to evaluate abnormal infants and red indicates sample of 30 clinical GMs used to evaluate normal infants. (Color figure online)

Figure 6 shows the power spectrum of the infant's limbs, which correspond to the limbs of the infant. Red indicates 30 clinical GMs that are used to evaluate the power spectrum of the normal infants, and green indicates 30 clinical GMs that are used to evaluate the power spectrum of abnormal infants. It can be seen that the normal infants have a higher power spectrum amplitude at the same frequency.

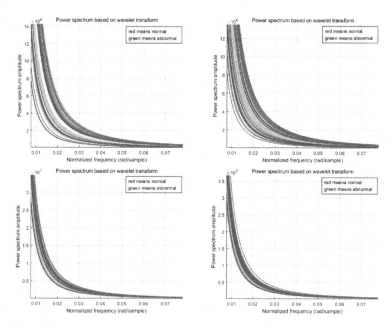

Fig. 6. Power spectrum of infant's limbs, where red indicates 30 clinical GMs used to evaluate power spectrum of normal infants, and green indicates 30 clinical GMs used to evaluate power spectrum of abnormal infants. (Color figure online)

3.4   Stacking Model Training

For the stacking model training, we combined the features of the left hand, right hand, left lower limb, right lower limb and the whole body as the fusion features. the stacking model training used both in the wavelet domain and power spectrum domain. Before the training, the method used a principal component analysis (PCA) to reduce the dimensionality. The principle of PCA is to project the original sample data into a new space and represent the original sample with fewer features. And then, we used the Support Vector Machine (SVM), Random forest (RF) [23], Adaboost as the first layer models, the second model is XGBoost. To prevent over-fitting in training, we used the 4-fold method to output the results of each part of the base model separately, and averaged the results as the input of the second layer.

4   Results

A true positive (TP) is defined as follows based on the literature review: infants with abnormal behavior are judged to be behaviorally abnormal. A false positive (FP) is a normal infant behavior judged as abnormal behavior. A true negative (TN) is a normal infant behavior determined to be normal behavior. A false negative (FN) is an infant with abnormal behavior judged to be behaving normally. Three metrics were defined, sensitivity (SE), specificity (SP), and accuracy (AC), as follows:

$$SE = \frac{TP}{TP + FN} \tag{5}$$

$$SP = \frac{TN}{TN + FP} \tag{6}$$

$$AC = \frac{TP + TN}{TP + FN + TN + FP} \tag{7}$$

The analysis results are listed in Table 1.

Table 1. Comparison of infant indicators under wavelet and power spectrum characteristics

Comparison of infant indicators under wavelet and power spectrum characteristics				
Feature	SE	SP	AC	TP/FN TN/FP
Wavelet	$(83.3 \pm 8.6)\%$	$(78.3 \pm 3.3)\%$	$(80.8 \pm 5.9)\%$	$12.5 \pm 1.29/2.5 \pm 1.29$ $11.75 \pm 0.5/3.25 \pm 0.5$
Power spectrum	$(93.3 \pm 6.6)\%$	$(85.0 \pm 10.0)\%$	$(89.2 \pm 8.3)\%$	$14.0 \pm 1.0/1.0 \pm 1.0$ $12.75 \pm 1.5/2.25 \pm 1.5$
Wavelet + power spectrum	$(95.0 \pm 5.0)\%$	$(91.7 \pm 6.3)\%$	$(93.3 \pm 5.7)\%$	$14.25 \pm 0.75/0.75 \pm 0.75$ $13.75 \pm 0.95/1.25 \pm 0.95$

In this paper, the final judgment is the weighted fusion result of power spectrum and wavelet models, the method used is a data adaptive weighted fusion estimation algorithm, according to the measured values obtained by each model, the optimal weighting factors corresponding to each model are searched adaptively, so that the fusion judgment value is optimized. The SE thus obtained was $(95.0 \pm 5.0)\%$, the SP was $(91.7 \pm 6.3)\%$, and the accuracy AC was $(93.3 \pm 5.7)\%$.

5   Conclusion

Because of whole body movements include activity from all segments of the cervical spinal cord to the lumbar spinal cord, the neural structures that produce GMs are most likely located in the spinal nerve center, since GMs first appears in the 9th week of pregnancy, participation of more advanced central structures above the brainstem is impossible. It is believed that the neural structure to produce whole body motion is the central pattern generator (CPG) which is located in the brainstem. CPG are neuronal circuits located in the spinal cord and brainstem that produce rhythmic movements such as walking, breathing, chewing and swimming. Prechtl [1] proposes that there are many non-rhythmic exercise patterns in human fetuses and young children (e.g. GM, fright, stretch, yawn), the neural mechanism in these cases should also be called CPG.

Affected by neurodevelopment, the characteristics of GMs are characterized by the fact that during the writhing phase (from full-term birth to 1 month of age), for normal babies, the whole body participates in the movement, giving a beautiful and complex impression, during the restless exercise phase (2–5 months after full-term birth), the abnormal GMs of an abnormal baby appear to be similar to normal restless movements, but are moderately or significantly exaggerated in terms of their range, speed, and unevenness, and exercise is uncoordinated rather than smooth. In this paper, the clinical observations of these forms of exercise are presented by parameters, the high or low value of these parameters reflects the clinically observed movement smooth or stiffness or the magnitude of acceleration changes, suggesting that infants with altered parameters may have brain development damage or abnormality.

This paper showed how the infant's left hand, right hand, left lower limb, right lower limb, and whole body can be tracked by KCF. Then, a wavelet waveform approximation analysis and wavelet power spectrum analysis can be performed on the five parts of the motion trajectory. In the wavelet waveform approximation analysis, the infant's wavelet waveform approximation map with abnormal behavior was more concentrated in a certain range than that of the infant with normal behavior, while the normal infant's motion waveform map was more scattered and irregular. In the wavelet power spectrum analysis, it was found that the power spectrum of the normal infant was higher at the same frequency. Stacking classification model training was applied to both the wavelet waveform approximation and wavelet power spectrum. A PCA dimension reduction was performed before the model training. Finally, the classification results are weighted and comprehensively judged to determine whether the infant's neurodevelopment is normal. Compared with the conclusions observed by doctors, the accuracy rate reached (93.3 ± 5.7)%.

References

1. Prechtl, H.F.R., Ferrari, F., Cioni, G.: Predictive value of general movements in asphyxiated fullterm infants. Early Human Dev. **35**, 91–120 (1993)
2. Adde, L., Helbostad, J.L., Jensenius, A.R., Taraldsen, G., Grunewaldt, K.H., Stoen, R.: Early prediction of cerebral palsy by computer-based video analysis of general movements: a feasibility study. Dev. Med. Child Neurol. **52**, 773–778 (2010)
3. Meinecke, L., Breitbach-Faller, N., Bartz, C., Damen, R., Rau, G., Disselhorst-Klug, C.: Movement analysis in the early detection of newborns at risk for developing spasticity due to infantile cerebral palsy. Hum. Mov. Sci. **25**, 125–144 (2005)
4. Adu, J., Gan, J., Wang, Y., Huang, J.: Image fusion based on nonsubsampled contourlet transform for infrared and visible light image. Infrared Phys. Technol. **61**, 94–100 (2013)
5. Henriques, J., Caseiro, R., Martins, P., Batista, J.: High-speed tracking with kernelized correlation filters. IEEE Trans. Pattern Anal. Mach. Intell. **37**, 583–596 (2015)
6. Rahmati, H., Martens, H., Aamo, O., Stavdahl, Ø., Støen, R., Adde, L.: Frequency analysis and feature reduction method for prediction of cerebral palsy in young infants. IEEE Trans. Neural Syst. Rehabil. Eng. **24**, 1225–1234 (2016)
7. Adde, L., Helbostad, J.L., Jensenius, A.R., Taraldsen, G., Støen, R.: Using computer-based video analysis in the study of fidgety movements. Early Human Dev. **85**, 541–547 (2009)

8. Fjørtoft, T., Einspieler, C., Adde, L., Strand, L.I.: Inter-observer reliability of the "Assessment of Motor Repertoire—3 to 5 Months" based on video recordings of infants. Early Human Dev. **85**, 297–302 (2009)

9. Adde, L., et al.: Early motor repertoire in very low birth weight infants in India is associated with motor development at one year. Eur. J. Paediatr. Neurol. **20**, 918–924 (2016)

10. Valle, S.C., Støen, R., Sæther, R., Jensenius, A.R., Adde, L.: Test–retest reliability of computer-based video analysis of general movements in healthy term-born infants. Early Human Dev. **91**, 555–558 (2015)

11. Stahl, A., Schellewald, C., Stavdahl, O., Aamo, O.M., Adde, L., Kirkerod, H.: An optical flow-based method to predict infantile cerebral palsy. IEEE Trans. Neural Syst. Rehabil. Eng. **20**, 605–614 (2012)

12. Yue, T., Suo, J., Cao, X., Dai, Q.: Efficient method for high-quality removal of nonuniform blur in the wavelet domain. IEEE Trans. Circuits Syst. Video Technol. **27**, 1869–1881 (2017)

13. Chai, Y., Li, H., Zhang, X.: Multifocus image fusion based on features contrast of multiscale products in nonsubsampled contourlet transform domain. Optik **123**, 569–581 (2012)

14. Uijlings, J.R.R., van de Sande, K.E.A., Gevers, T., Smeulders, A.W.M.: Selective search for object recognition. Int. J. Comput. Vis. **104**, 154–171 (2013)

15. Patel, M., Lal, S., Kavanagh, D., Rossiter, P.: Fatigue detection using computer vision. Int. J. Electron. Telecommun. **56**, 457–461 (2010)

16. Kuen, J., Lim, K.M., Lee, C.P.: Self-taught learning of a deep invariant representation for visual tracking via temporal slowness principle. Pattern Recogn. **48**, 2964–2982 (2015)

17. Lin, B., Wei, X., Junjie, Z.: Automatic recognition and classification of multi-channel microseismic waveform based on DCNN and SVM. Comput. Geosci. **123**, 111–120 (2019)

18. Ojala, T., Pietikäinen, M., Harwood, D.: A comparative study of texture measures with classification based on featured distributions. Pattern Recogn. **29**, 51–59 (1996)

19. Afifi, S., GholamHosseini, H., Sinha, R.: A system on chip for melanoma detection using FPGA-based SVM classifier. Microprocess. Microsyst. **65**, 57–68 (2019)

20. Xu, J., Tang, Y.Y., Zou, B., Xu, Z., Li, L., Lu, Y.: The generalization ability of online SVM classification based on Markov sampling. IEEE Trans. Neural Netw. Learn. Syst. **26**, 628–639 (2015)

21. Xiao, J.: SVM and KNN ensemble learning for traffic incident detection. Physica A: Stat. Mech. Appl. **517**, 29–35 (2019)

22. NaNa, Z., Jin, Z.: Optimization of face tracking based on KCF and Camshift. Proc. Comput. Sci. **131**, 158–166 (2018)

23. Nedjar, I., Daho, M., Settouti, N., Mahmoudi, S., Chikh, M.: RANDOM forest based classification of medical x-ray images using a genetic algorithm for feature selection. J. Mech. Med. Biol. **15**, 1540025 (2013)

24. Fern, A., Schapire, R.: Online ensemble learning: an empirical study. Mach. Learn. **53**, 71–109 (2003)

Detecting Neurodegenerative Disease from MRI: A Brief Review on a Deep Learning Perspective

Manan Binth Taj Noor[1], Nusrat Zerin Zenia[1], M. Shamim Kaiser[1(✉)] [iD],
Mufti Mahmud[2(✉)] [iD], and Shamim Al Mamun[1] [iD]

[1] Institute of Information Technology, Jahangirnagar University, Dhaka, Bangladesh
{manan.noor,nusratzenia,mskaiser,shamim}@juniv.edu
[2] Department of Computing and Technology, Nottingham Trent University,
Clifton, Nottingham MG11 8NS, UK
mufti.mahmud@ntu.ac.uk, mufti.mahmud@gmail.com

Abstract. Rapid development of high speed computing devices and infrastructure along with improved understanding of deep machine learning techniques during the last decade have opened up possibilities for advanced analysis of neuroimaging data. Using those computing tools Neuroscientists now can identify Neurodegenerative diseases from neuroimaging data. Due to the similarities in disease phenotypes, accurate detection of such disorders from neuroimaging data is very challenging. In this article, we have reviewed the methodological research papers proposing to detect neurodegenerative diseases using deep machine learning techniques only from MRI data. The results show that deep learning based techniques can detect the level of disorder with relatively high accuracy. Towards the end, current challenges are reviewed and some possible future research directions are provided.

Keywords: Machine learning · Alzheimer's disease · Schizophrenia · Parkinson's disease · MRI

1 Introduction

With the increasing necessity for delivering improved healthcare services, extensive research has been carried out during the last decade towards developing novel computational tools for the Neuroscientists to diagnose disease and devise appropriate treatment strategies [21,22,28]. This remarkable progress, achieved through a multi-disciplinary approach, has contributed in an improved understanding of human brain and its inter-connectivity. Especially, much progress has been made in the classification of Neurodegenerative diseases (ND) using machine learning which is also reflected in the current research trends of ND classification using deep learning (DL) methods as seen in Fig. 1. The annual trend data were obtained by googling with search terms consisting of three

© Springer Nature Switzerland AG 2019
P. Liang et al. (Eds.): BI 2019, LNAI 11976, pp. 115–125, 2019.
https://doi.org/10.1007/978-3-030-37078-7_12

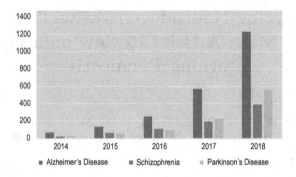

Fig. 1. Research trends of Alzheimer's disease, Schizophrenia, Parkinson's diseases classification using DL.

keywords– one of the diseases (i.e., either "Alzheimer's" or "Parkinson's" or "Schizophrenia"), "DL", and "MRI" for years ranging from 2014 to 2019.

Neurodegenerative diseases such as Parkinson's disease, Alzheimer's disease etc. are characterised by the retardation of regular operations of brain functions. A variety of DL methods have been proposed for ND detection and classification of the diseases including their stage from a number of different neuroimaging techniques such as MRI, CT, PET, etc. Due to the increasing number of available approaches to analyse these data, it is imperative to review the existing ones to be able to select an appropriate technique for a given dataset. There have been several review works in multiple directions where different approaches were made to synthesize the applications of machine learning and big data to study mental health [34]. Various machine learning based tasks have been explored on connectome data from MRI which aims to better diagnose neurological disorders [4]. In [10] authors have investigated the application of DL to better understand and diagnose Parkinson's disease. A detailed survey on DL applied to the analysis of various medical image such as neuro, pulmonary, pathology etc. has been conducted in [18]. The ultimate objective of this research is to put forth an overview of the DL techniques to detect ND from MRI and its variants. However, being aware of space restrictions, only a brief review was performed and reported in this paper.

2   Deep Learning

DL- being a machine learning method - can be used to build models which learn high dimensional features from data. A number of DL structures such as Convolutional Neural Network (CNN), Deep Neural Network (DNN), Recurrent Neural Network (RNN), Autoencoder (AE), Deep Belief Network (DBN), Probabilistic Neural Network (PNN), etc. have been reported in the literature. These structures can be used to classify various ND with very high accuracy [20]. Figure 2 represents a summary of different MRI techniques and concerned DL methods covered in this research.

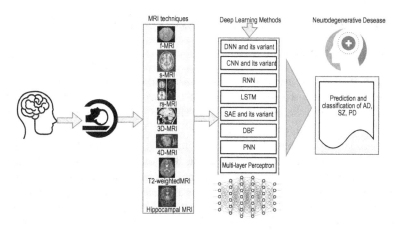

Fig. 2. Applications of DL to classify and predict neurodegenerative disease based on different variants of MRI

A neural network having many hidden layers is generally referred as DNN. In deep feed forward neural networks also known as Multi-layered Network (MLN), information can travel only one way (forward) with no feedback in the network; whereas, feedback neural network has some kind of internal recurrence and hence feed back to a neuron or layer that has already received and processed that signal [29].

CNN uses convolution operation in place of simple matrix multiplication in at least one of their layers. It is mainly used in an unstructured data set (e.g. image, video). 2D-CNN is able to leverage features from only spatial dimensions (height and width) whereas 3D-CNN can preserve temporal dimensions by predicting volumetric patch of neuroimaging data [32].

RNN is known as memory network which depends on present input and recent previous computations to produce the output of new information. RNN is mostly used in Natural Language Processing [6].

An AE learns to compress its input to its output in an unsupervised manner. It has a hidden layer which is used to describe a code to constitute the input. An adaption of AE incorporating DL architecture is Stacked autoencoder or StAE. StAE is a neural network with a stacked number (i.e., multiple) AE layers of AE [26].

DBN is mainly constituted of a stack of restricted Boltzmann machines (RBM) where each RBM layer needs to communicate with both the foregoing and successive layers. Deep-belief networks are used to identify, group and originate images, video clips and motion-capture data [27].

In PNN, the Probability Distribution Function (PDF) is used to compute the probability of a new input data. Finally, Bayes rule is used to assign the new input data with the class that has the highest posterior probability [23].

3   Identifying Neurodegenerative Diseases

3.1   Alzheimer's Disease

AD is characterized by escalating mental degradation that generally occur in middle or old age, due to deterioration of the brain. However, research efforts are being conducted to correctly discover the cause and way to detect the AD patterns from neuroimages in an automated way. Following that authors in [26] present deep artificial neural networks consisting of sparse autoencoders and convolutional neural networks by applying 3D convolutions on the whole MRI image to discriminate between healthy controls (HC) and diseased brains.

Application of deep CNN has been very common in detecting AD as seen in [32] where there has been demonstrated CNN architecture by extracting scale and shift invariant low to high-level features to differentiate HC and AD in older adults. Similarly authors in [33] applied CNN DL architecture (LeNet) to detect AD from HC. Authors in [12] came up with a novel and faster framework by using hyper-parameters from a very deep image classifier CNN using small medical image dataset to diagnose AD. Farooq et al. also used CNN framework for multiclass classification among AD, prodromal stages of AD and HC [9]. Furthermore, Spasov et al. presented a parameter efficient 3D separable CNN architecture to predict mild cognitive impairment (MCI) to AD conversion along with classification of AD and HC [37]. Böhle et al. classified AD and HC using layer-wise relevance propagation of CNN on MRI data [3]. Basaia et al. was also found to use CNN to distinguish among AD, MCI conversion to AD and stable MCI based on a single cross-sectional brain MRI scan [2]. In addition, Ullah et al. came up with CNN architecture to detect AD and Dementia from 3D MRI image with manual feature extraction [39].

Amoroso et al. proposed a pure machine learning approach exploiting Random Forest for feature selection and a DNN for classification to early detect Alzheimer's disease [1]. A Long short term memory (LSTM) autoencoder has been approached in [17] consisting of RNNs to characterize the early prediction of MCI progression to AD.

Luo et al. provided an automatic AD detection algorithm using CNN on 3D brain MRI in which the 3D topology of whole brain is considered [19]. Dolph et al. tried first application of a model consisting of Stacked auto-encoder (SAE)-DNN based multi-classifier that can learn complex nonlinear atrophy patterns for classification of AD, MCI, and Cognitive Normal (NC) using both in-house and public-domain standardized CADDementia framework [7]. Bäckström et al. proposed three-dimensional convolutional (3D ConvNet) network for automatically learning the features and detecting AD on a by relatively pre-processing and fine tuning a large size MRI dataset using 3D ConvNet hyperparameters [5].

A statistical feature Gray Level Cooccurrence (GLCM) based model exploiting principal component analysis (PCA) and finally PNN for training and classification has been proposed by Mathew et al. to classify AD, MCI and NC [23].

3.2   Schizophrenia

Schizophrenia (SZ) is a major psychiatric disorder associated with structural and functional brain anomalies that gradually results in cognition, emotion, and behavioral impairments. In recent years, many researchers have concentrated to develop automated diagnosis techniques of SZ using DL and MRI data. Qureshi et al. has proposed 3D-CNN based DL classification utilizing independent component analysis (ICA) to semi-automatically separate the noise and artefacts from rs-fMRI with a view to distinguishing patients with SZ and HC [31]. Instead of 3D-CNN, 2D-CNN with ICA has been applied on rs-fMRI in [30] for achieving the same purpose.

However, a majority of reported approaches have applied DNN based classification technique to diagnose SZ [24,29,38,42]. Srinivasagopalan et al. proves that DL can be a paradigm shift to detect SZ by comparing DNN based classification results with that of conventional machine learning approach (e.g. Support vector machine (SVM), logistic regression etc.) [38]. Matsubara et al. proposed a deep neural generative model (DGM) of rs-fMRI data. Along with DNN, DGM has applied a dedicated approach to evaluate the contribution weight of different brain region to the diagnosis using Bayes' rule [24]. Besides, DNN and layer-wise relevance propagation (LRP) is used to improve the classification accuracy of SZ patients [42]. In addition, DNN-based multi-view models comprising deep canonical correlation analysis (DCCA), deep canonically correlated auto-encoders (DCCAE) and SVM with Gaussian kernel is used to determine SZ in [29]. It is noted that multimodal features are mainly considered as parameter for DNN based classification.

Feed forward neural network is applied to analyze normal and SZ subjects from multisite S-MRI data in [40]. In contrast, Han et al. concept was to utilize the benefit of back propagation with feed forward on rs-fMRI features to distinguish SZ [11].

Attempt has been made to explore the performance of DBN in case of discriminating the normal and SZ subjects by taking region of interest (ROI) and morpho-metry data into consideration in [16] and [27] respectively. Latha et al. has used Stochastic gradient descent (SGD), adaptive gradient (Adagrad) and root-mean-square propagation (RMSProp) to train the considered region [16]. On the other hand, multivariate analysis was done for visualizing the most affected brain regions in [27]. Moreover, some current studies have employed auto-encoder for functional connectivity (FC) feature extraction [13,25,43]. Thereafter, these trained features are applied to SVM classifier [25] or DNN classifier for automatic diagnosis of individuals with SZ [13,43]. Apart from that Dakka et al. has successfully demonstrated the feasibility of R-CNN involving a 3-D CNN with long short term memory (LSTM) units [6].

3.3   Parkinson's Disease

PD comes in as a result of movement disorder. DL has also been incorporated to demonstrate it's signs and detection from neuro images. Following that Kollias et al. proposed a deep neural architecture including CNN deriving rich internal depiction from input data and B-LSTM/GRU RNNs to analyse time progression of the inputs for delivering the final predictions [15]. Then, Shinde et al. proposed to differentiate PD from HC by employing a fully automated CNN with discriminative localization architecture for creating prognostic and diagnostic biomarkers of PD from Neuromelanin sensitive MRI [35]. On the other hand Kollia et al. used a convolutional-RNN for PD prediction through extraction of latent variable information from trained deep neural networks using both MRI and DaT Scan data [14].

Moreover, Sivaranjini et al. contributed in analysing T2 weighted MR brain images to classify between HC and PD by applying deep CNN architecture AlexNet [36].

Esmaeilzadeh et al. used a 3D Convolutional Neural Network incorporating a voxel-based approach for brain image segmentation extracting data augmentation techniques to expand the training set size to classify PD and HC [8].

Table 1 presents a summary of mentioned DL applications including the type of MRI, dataset and classification accuracy.

4   Dataset

Most of the work done for Alzheimer's Disease Neuroimaging Initiative has used MCI and AD datasets from ADNI data repository. The dataset includes demographic information, raw neuroimaging scan data, APOE genotype, CSF measurements, neuropsychological test scores, and diagnostic information. There has also been used OASIS dataset which consists of longitudinal neuroimaging, clinical, cognitive, and biomarker data for normal aging and Alzheimer's Disease.

Center for Biomedical Research Excellence (COBRE) dataset is prevalent in schizophrenia related studies. The dataset includes details of 72 subjects (age range: 18–65) from both the HC and SZ groups. Some studies have used OpenfMRI database which contains the information about 50 SZ patients, 49 patients with bipolar disorder, and 122 HC. Another SZ dataset is function biomedical informatics research network (fBIRN) which contains 135 SZ patients (including SZ and schizoaffective disorder) and 169 HC. Moreover, several studies has been found while reviewing that have collected data from multiple sites or different hospitals to validate their proposed model.

Kollias et al. used the NTUA Parkinson dataset which consists of MRI, DaT Scans and clinical data of 55 patients with PD and 23 subjects with PD-related syndromes. Some studies have also used the Parkinson's Progression Markers Initiative (PPMI) public domain database to detect biomarkers of PD progression. Another study in [35] acquired MR images of subjects from Department of Neurology, National Institute of Mental Health and Neurosciences (NIMHANS), Bangalore, India.

Table 1. Summary of DL based studies for prediction and classification of AD, SZ and PD

Reference	Data type	Dataset	DL technique	Performance
Alzheimer				
[26]	MRI	ADNI	SAE-3D CNN	2D-85.53% 3D-89.47%
[32]	MRI, fMRI	ADNI	CNN	fMRI-99.9%, MRI-98.84%
[33]	fMRI	ADNI	CNN	96.86%
[12]	MRI	OASIS	CNN	73.75%
[9]	MRI	ADNI	CNN	98.8%
[1]	MRI	ADNI	DNN	38.8%
[37]	sMRI	ADNI	3D-CNN	100%
[3]	sMRI	ADNI	CNN	
[2]	sMRI	ADNI	CNN	99%
[17]	Hippocampal MRI	ADNI	LSTM-RNNs	
[19]	3D MRI	ADNI	CNN	
[7]	sMRI	ADNI	SAE-DNN	≈(51–58.0)%
[5]	MRI	ADNI	3D ConvNet	98.74%
[23]	fMRI	ADNI	PNN	67.50%
[39]	3D MRI	OASIS	CNN	80.25%
Schizophrenia				
[40]	sMRI	Multisite	Feed Forward NN	
[13]	rs-fMRI	COBRE	SAE-DNN	
[29]	MRI	COBRE	DNN	
[27]	sMRI	Multisite	DBN	73.6%
[25]	fMRI	COBRE	SAE	92%
[42]	rs-fMRI	Multisite	DNN	84.75%
[11]	rs-fMRI	Hospital	DNN	79.3%
[6]	4D-fMRI	fBIRN	RNN	≈63%
[41]	sMRI, fMRI	fBIRN	MLP	
[16]	MRI	COBRE	DBN	90%
[43]	fc-MRI	Multisite, COBRE	DDAE	(81–85)%
[24]	rs-fMRI	OpenfMRI	DNN	0.766
[38]	fMRI, sMRI	COBRE	DNN	94.4%
[30]	rs-fMRI		2D-CNN	
[31]	rs-fMRI	COBRE	3D-CNN	98.09%
Parkinson				
[15]	MRI, DAT Scan	NTUA	CNN-RNN	98%
[8]	MRI	PPMI	3D-CNN	100%
[35]	sMRI	NIMHANS	CNN	85.7%
[14]	sMRI, DaT Scan	NTUA	CNN-RNN	98%
[36]	T2 weighted MRI	PPMI	CNN	88.9%

5   Performance Analysis

All the referred studies included in this paper incorporate several aspects of work from AD prediction, MCI to AD conversion, multiclass AD classification etc. Authors in [37] achieved highest accuracy of about 100% using 3D-CNN for classifying AD from HC. But an accuracy of 38.8% has been achieved which featured a scientific challenge placing third over 19 participating teams to classify AD which is comparatively lower than all other [1].

Highest accuracy 98.09% of schizophrenia detection has been observed in [31] which have employed 3D-CNN based classification. Above 90% accuracy is shown in [38]vand [16]. The other studies perceived the accuracy ranges from 70–80%.

By using 3D CNN [8] achieved 100% accuracy on the validation and test sets for PD diagnosis. At the same time, study in [35] discriminated PD from typical parkinsonian syndromes having 85.7% test accuracy.

6   Challenges and Future Perspective

Rapid development in the computing capabilities and better understanding of DL algorithms enable us to use DL based neurodegenerative disease prediction framework. However, a number of challenges in the existing techniques require further research in order to improve the overall accuracy of the system. Some of these are outlined below:

The supervised architecture is limited due to huge effort for creating label data, low scalability and selection of appropriate bias level. Unsupervised learning is not an usual option to be considered for image analysis. However, unsupervised architecture not only learn features from the dataset, but also design data-driven decision support system from these data. Thus unsupervised deep architecture can be used in solving medical imaging related problems.

In real-time applications, collected imaging data are big data and anomaly events occur rarely, thus collected data are imbalanced. Thus the classifier will be biased towards majority classes and will provide low accuracy. Further research may be required to improve data imbalances.

To handle real time imaging data, stream processing may be required which might demand changes in the parallel computing algorithm.

7   Conclusion

DL and Brain research has a major emphasis on the current research trends. This paper, briefly yet comprehensively reviewed different studies which aim to identify three prominent neurodegenerative disorders (i.e., Alzheimer's Disease, Schizophrenia, and Parkinson's Disease) by applying DL-based techniques only from various MRI data. Towards the end, the related datasets and performances of each method are summarized. Finally, the gaps in the research area along with suggestion of future trend are mentioned.

References

1. Amoroso, N., et al.: Deep learning reveals Alzheimer's disease onset in MCI subjects: results from an international challenge. J. Neurosci. Methods **302**, 3–9 (2018)
2. Basaia, S., et al.: Automated classification of Alzheimer's disease and mild cognitive impairment using a single MRI and DNN. Neuroimage Clin. **21**, 101645 (2019)
3. Bohle, M.: Layer-wise relevance propagation for explaining DNN decisions in MRI-based Alzheimer's disease classification. Front. Aging Neurosci. **11**, 194 (2019)
4. Brown, C.J., Hamarneh, G.: Machine learning on human connectome data from MRI. CoRR abs/1611.08699 (2016)
5. Bäckström, K., et al.: An efficient 3D deep convolutional network for Alzheimer's disease diagnosis using MR images. In: Proceedings of the ISBI 2018, pp. 149–153 (2018)
6. Dakka, J., et al.: Learning neural markers of schizophrenia disorder using recurrent neural networks. CoRR abs/1712.00512 (2017)
7. Dolph, C.V., Alam, M., Shboul, Z., Samad, M.D., Iftekharuddin, K.M.: Deep learning of texture and structural features for multiclass Alzheimer's disease classification. In: Proceedings of the IJCNN 2017, pp. 2259–2266 (2017)
8. Esmaeilzadeh, S., Yang, Y., Adeli, E.: End-to-end Parkinson disease diagnosis using brain MR-images by 3D-CNN. CoRR abs/1806.05233 (2018)
9. Farooq, A., Anwar, S., Awais, M., Rehman, S.: A deep CNN based multi-class classification of Alzheimer's disease using MRI. In: Proceedings of the IEEE IST 2017, pp. 1–6 (2017)
10. Gottapu, R.D., Dagli, C.H.: Analysis of Parkinson's disease data. Proc. Comput. Sci. **140**, 334–341 (2018)
11. Han, S., et al.: Recognition of early-onset schizophrenia using deep-learning method. Appl. Inform. **4**(1), 16 (2017)
12. Islam, J., Zhang, Y.: A novel deep learning based multi-class classification method for Alzheimer's disease detection using brain MRI data. In: Zeng, Y., et al. (eds.) BI 2017. LNCS, vol. 10654, pp. 213–222. Springer, Cham (2017). https://doi.org/10.1007/978-3-319-70772-3_20
13. Kim, J., et al.: Deep NN with weight sparsity control and pre-training extracts hierarchical features and enhances classification performance: evidence from whole-brain resting-state functional connectivity patterns of schizophrenia. NeuroImage **124**, 127–146 (2015)
14. Kollia, I., Stafylopatis, A., Kollias, S.D.: Predicting Parkinson's disease using latent information extracted from deep neural networks. CoRR abs/1901.07822 (2019)
15. Kollias, D., et al.: Deep neural architectures for prediction in healthcare. Complex Intell. Syst. **4**(2), 119–131 (2018)
16. Latha, M., Kavitha, G.: Detection of Schizophrenia in brain MR images based on segmented ventricle region and DBNs. Neural Comput. Appl. **31**, 5195–5206 (2018)
17. Li, H., Fan, Y.: Early prediction of Alzheimer's disease dementia based on baseline hippocampal MRI and 1-year follow-up cognitive measures using deep recurrent neural networks. CoRR abs/1901.01451 (2019)
18. Litjens, G., et al.: A survey on deep learning in medical image analysis. Med. Image Anal. **42**, 60–88 (2017)
19. Luo, S., Li, X., Li, J.: Automatic Alzheimer's disease recognition from MRI data using deep learning method. J. Appl. Math. Phys. **05**, 1892–1898 (2017)

20. Mahmud, M., Kaiser, M.S., Hussain, A., Vassanelli, S.: Applications of deep learning and reinforcement learning to biological data. IEEE Trans. Neural Netw. Learn. Syst **29**(6), 2063–2079 (2018)
21. Mahmud, M., Vassanelli, S.: Processing and analysis of multichannel extracellular neuronal signals: state-of-the-art and challenges. Front. Neurosci. **10**(JUN), 248 (2016). https://doi.org/10.3389/fnins.2016.00248
22. Mahmud, M., Vassanelli, S.: Open-source tools for processing and analysis of in vitro extracellular neuronal signals. In: Chiappalone, M., Pasquale, V., Frega, M. (eds.) In Vitro Neuronal Networks. AN, vol. 22, pp. 233–250. Springer, Cham (2019). https://doi.org/10.1007/978-3-030-11135-9_10
23. Mathew, N.A., Vivek, R.S., Anurenjan, P.R.: Early diagnosis of Alzheimer's disease from MRI images using PNN. In: Proceedings of the IC4, pp. 161–164 (2018)
24. Matsubara, T., et al.: Deep neural generative model of functional MRI images for psychiatric disorder diagnosis. IEEE Trans. Biomed. Eng. **66**(10), 2768–79 (2019)
25. Patel, P., Aggarwal, P., Gupta, A.: Classification of schizophrenia versus normal subjects using deep learning. In: Proceedings of the ICVGIP, India, pp. 281–286 (2016)
26. Payan, A., Montana, G.: Predicting Alzheimer's disease: a neuroimaging study with 3D convolutional neural networks. CoRR abs/1502.02506 (2015)
27. Pinaya, W.H., et al.: Using deep belief network modelling to characterize differences in brain morphometry in schizophrenia. Sci. Rep. **6**, 38897 (2016)
28. Poldrack, R., et al.: Computational and informatic advances for reproducible data analysis in neuroimaging. Annu. Rev. Biomed. Data Sci. **2**, 119–138 (2019)
29. Qi, J., Tejedor, J.: Deep multi-view representation learning for multi-modal features of the schizophrenia and schizo-affective disorder. In: Proceedings of the IEEE ICASSP, pp. 952–956 (2016)
30. Qiu, Y., et al.: Classification of schizophrenia patients and healthy controls using ICA of complex-valued fMRI data and convolutional neural networks. In: Lu, H., Tang, H., Wang, Z. (eds.) ISNN 2019. LNCS, vol. 11555, pp. 540–547. Springer, Cham (2019). https://doi.org/10.1007/978-3-030-22808-8_53
31. Qureshi, M.N.I., Oh, J., Lee, B.: 3D-CNN based discrimination of schizophrenia using resting-state fMRI. Artif. Intell. Med. **98**, 10–17 (2019)
32. Sarraf, S., Tofighi, G.: DeepAD: Alzheimer's disease classification via deep convolutional neural networks using MRI and fMRI. bioRxiv (2016)
33. Sarraf, S., Tofighi, G.: Classification of Alzheimer's disease using fMRI data and deep learning CNNs. CoRR abs/1603.08631 (2016)
34. Shatte, A., Hutchinson, D., Teague, S.: Machine learning in mental health: a scoping review of methods and applications. Psychol. Med. **49**, 1–23 (2019)
35. Shinde, S., et al.: Predictive markers for Parkinson's disease using deep neural nets on neuromelanin sensitive MRI. NeuroImage: Clin. **22**, 101748 (2019)
36. Sivaranjini, S., Sujatha, C.M.: Deep learning based diagnosis of Parkinson's disease using convolutional neural network. Multimed. Tools Appl. (2019)
37. Spasov, S., et al.: A parameter-efficient DL approach to predict conversion from mild cognitive impairment to Alzheimer's disease. NeuroImage **189**, 276–287 (2019)
38. Srinivasagopalan, S., et al.: A deep learning approach for diagnosing schizophrenic patients. J. Exp. Theoret. Artif. Intell. **31**, 1–14 (2019)
39. Ullah, H.M.T., et al.: Alzheimer's disease and dementia detection from 3D brain MRI data using deep CNNs. In: Proceedings of the I2CT 2018, pp. 1–3 (2018)
40. Ulloa, A., et al.: Synthetic structural magnetic resonance image generator improves deep learning prediction of schizophrenia. In: Proceedings of the IEEE MLSP, pp. 1–6 (2015)

41. Ulloa, A., Plis, S.M., Calhoun, V.D.: Improving classification rate of schizophrenia using a multimodal multi-layer perceptron model with structural and functional MR. CoRR abs/1804.04591 (2018)
42. Yan, W., et al.: Discriminating schizophrenia from normal controls using resting state functional network connectivity: a deep neural network and layer-wise relevance propagation method. In: Proceedings of the MLSP, pp. 1–6 (2017)
43. Zeng, L.L., et al.: Multi-site diagnostic classification of schizophrenia using discriminant deep learning with functional connectivity MRI. EBioMedicine **30**, 74–85 (2018)

The Changes of Brain Networks Topology in Graph Theory of rt-fMRI Emotion Self-regulation Training

Lulu Hu[1], Qiang Yang[1], Hui Gao[1], Zhonglin Li[2], Haibing Bu[1],
Bin Yan[1], and Li Tong[1(✉)]

[1] National Digital Switching System Engineering
and Technological Research Center, Zhengzhou, Henan, China
tttocean@163.com
[2] Henan Provincial People's Hospital, Zhengzhou, Henan, China

Abstract. Neural feedback technology based on rt-fMRI (real-time functional Magnetic Resonance Imaging) provides a new non-invasive method to improve the cognitive function of the human brain, which achieves by training the human brain to regulate emotion. At the same time, brain network approaches based on graph theory is a hot spot. In this paper, we focus on the changes in the human brain's small-world topology and network efficiency in graph theory before and after neurofeedback experiments. We designed an emotion self-regulation training with rt-fMRI, and acquired data from 20 participants, divided into the experimental group (EG) and the control group (CG). Subsequently, we constructed the brain network through the Anatomic Automatic Labelling (AAL) atlas, compared the topological changes of brain network between the EG and the CG in emotion self-regulation training. Our results show that both the EG and the CG have small-world topology, there are differences in small-world topology with emotion self-regulation training. Additionally, local efficiency is significantly different under certain sparsity, which suggests that emotional regulation has a positive effect on local networks. However, there is no significant difference in global efficiency, which means that the global network property does not change in emotion regulation training.

Keywords: Graph theory · Brain network · Emotion self-regulation training · rt-fMRI · Small-world

1 Introduction

From the perspective of the whole brain network, thoroughly researching the mechanism of emotional real-time neurofeedback regulation training is of great significance for humans to independently regulate brain activity, and can be promoted from the basic science field to clinical application to achieve an emotionally regulation for emotional disorder patient. The human brain is organized into a complex network of structures and functions for the separation and integration of information transmission. How to effectively reveal the neural mechanism of brain structures and functions is very important in brain science research. In recent years, complex brain network

© Springer Nature Switzerland AG 2019
P. Liang et al. (Eds.): BI 2019, LNAI 11976, pp. 126–135, 2019.
https://doi.org/10.1007/978-3-030-37078-7_13

analysis methods based on graph theory have been widely used. Brain network analysis based on graph theory provides a new perspective for assessing the symptoms of patients with emotional disorders. Graph theory is a mathematical technique which can model the brain as a complex network represented graphically by sets of nodes and edges [1, 2]. The brain network is represented as a directed or undirected graph by the connection of the edges and the nodes. In a network, some key global network metrics are: small-world [3], rich-club [4], and network efficiency and so on; the nodal properties can be measured by some metrics such as the degree, efficiency, and betweenness centrality [5]. Mathematically, these metrics can be well characterized by "graphs".

In the past years, many outstanding achievements have been made in brain network approaches. The small-world topology of the human brain function network was first discovered by Salvador [6] who used the resting-state fMRI technology, and was verified repeatedly in later studies. van den Heuvel et al. [7] indicated a strong positive association between the global efficiency of the functional brain networks and intellectual performance by resting-state data. In addition, in the field of clinical neuroscience, graph theory analysis has also received great attention. Ye et al. [8] showed that compared with healthy controls, MDD patients showed higher local efficiency and modularity. Zhang et al. [9] used the data of the resting-state brain network to construct a complex brain network of the patients with depression, and carried out a more detailed analysis.

These studies indicated that brain network analysis based on graph can effectively reflect the functional status of the brain. Therefore, we used the graph theory approach to compare the topological metrics differences of the different groups in the emotion self-regulation training to characterize the functional integration and separation, information transmission efficiency of the brain network under the neural feedback, in order to evaluate the neurofeedback effect from a new perspective. The small-world topology in the emotion regulation training shows that the brain network of people after neurofeedback training can continuously integrate information from different brain regions. This paper shows that local information transmission efficiency has changed in some specific brain regions. However, discuss from global level, the efficiency of brain network information transmission after neurofeedback is preserved.

2 Methods

2.1 Participants

20 healthy volunteers (14 males with mean age of 23.4 years, standard deviation (SD) = 2.6; 6 females with mean age of 24, SD = 2) were recruited from China National Digital Switching System Engineering and Technological Research Center. All participants were right-handed without a history of neurological or psychiatrical diseases. Before the experiment, all subjects were informed of the experimental content and signed informed consent. The research protocol was approved by the ethics committee of Henan Provincial People's Hospital, and the participants were given certain financial compensation after the experiment.

2.2   Experimental Paradigm

Before one or two days of the experiment, the subjects were trained about 30 min and were informed of the purpose of the experiment so that they were familiar with the entire experimental procedure. The neurofeedback training experiment consists of five stages, the specific design is shown in Fig. 1. A 320-second resting-state scan was performed before and after the neurofeedback training (stage 1 and stage 5). At stage 2, the subjects recalled alternately the happy memories and sad memories prepared in advance, inducing the corresponding emotional state. At stage 3, the classifier model is trained through the imaginary data collected in stage 2. Firstly, the collected data is pre-processing, through general liner model (GLM) in SPM, the brain regions activation can be obtained during the recall of happy and sad memories during imaginary training. Then the feature template is extracted to create a sample by RFE method. Finally, the created samples are input into the SVM classifier to train classification model, which is used to identify the real-time emotional state and generate feedback information in the neural feedback training. At stage 4, the participants were arranged to three feedbacks of the neurofeedback training. The subjects should try to increase the bar on the left side of the feedback screen by recalling the happy memories, they can adjust the strategy by changing the strips. The change in the height of the strip is determined by the emotional state recognized by the classifier and the distance of the current state from the classification surface. Feedback display screen 1(DS1) is the initial state of feedback to the subject. When the accumulated state is higher than the initial state, the left bar becomes red, that is, feedback display screen 2(DS2); when the accumulated state is lower than the initial state, the left bar turns green, that is, feedback display screen 3(DS3). During the neurofeedback training, the experimental group feedback the information of the current emotion state, while the control group did not feedback any information.

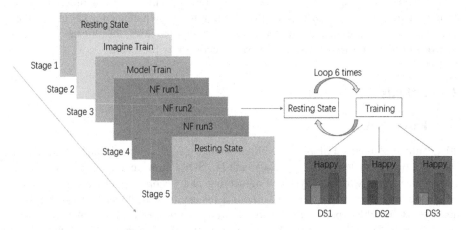

Fig. 1. Neural feedback experiment flow chart (Color figure online)

2.3 Data Acquisition

The experiment data were acquired from the earlier published research, more details can be found in the paper Part 3 Sect. 2 [10].

2.4 fMRI Data Preprocessing

The fMRI resting-state data pre-processing was under DPABI [11] (http://www.restfmri.net), divided into seven steps. (1) Remove the first ten volumes as a result of magnetic equilibration effects and participants were not calm; (2) slice timing, with the middle slice as the reference slice; (3) realign, select the data with the maximum head motion translation not exceeding 3 mm; (4) space normalization, standardize the data space to the MNI standard human brain template, with the voxel size $3 \times 3 \times 3$ re-sampling; (5) smoothing, using a Gaussian kernel function with a height and width of 8 mm; (6) filtering, bandwidth of 0.01–0.08 Hz; (7) regression of noise signals of all images, remove the impact of interference factors of 6 head motion parameters, whole brain average signal, CSF, WM, etc.

2.5 Construction of Functional Brain Network

According to the Anatomic Automatic Labelling (AAL) [12] template, the brain is divided into 90 regions. The time series of 90 brain regions of each subject was extracted by GRETNA [13] (http://www.nitrc.org/projects/gretna) which is based on the MATLAB platform, and the whole brain functional connection network of each subject was constructed to obtain a correlation coefficient matrix (90×90), then perform a Fisher transform to convert the correlation coefficient to Z value, making the matrix closer to a normal distribution. After the correlation coefficient matrix of each subject is obtained then the matrix is binarized, which is represented by 1 between the connected nodes, and 0 is not connected to represent the brain network. Because there is no authoritative method of setting the sparsity, we have ensured that the brain network has both significant connections and the connection is not very dense. the sparsity range between 0.1 and 0.5 is selected.

2.6 Global Network Parameters

The global network parameters calculated in this paper include clustering coefficient, shortest path length, global efficiency, and local efficiency.

Clustering Coefficient
The clustering coefficient describes the proportion of neighbors of nodes in the network which are also adjacent to each other. For the i-th node, the clustering coefficient is defined as:

$$C_i = \frac{E_i}{\frac{1}{2} k_i(k_i - 1)} \tag{1}$$

E_i is the number of the actual edges of the neighbors, k_i is defined as the number of other nodes in the network that are directly connected to node i.

Shortest Path Length

The shortest path length describes the shortest number of connected edges between two nodes. The calculation formula is:

$$L = \frac{1}{N(N-1)} \sum_{i \neq j} d_{ij} \tag{2}$$

N represents the number of nodes, and d_{ij} represents the distance between node i and node j.

Small-World Topology

The regular network has a higher shortest feature path length and a higher clustering coefficient. The random network has a shorter shortest feature path length and a lower clustering coefficient. A small-world network is a network topology between a regular network and a random network. It has high clustering coefficient and shortest path length. The formula is:

$$\gamma = \frac{C_{real}}{C_{random}} \gg 1 \tag{3}$$

$$\lambda = \frac{L_{real}}{L_{random}} \approx 1 \tag{4}$$

$$\sigma = \frac{\gamma}{\lambda} \tag{5}$$

C_{real} is the clustering coefficient of the real network, C_{random} is the clustering coefficient of the random network; L_{real} is the shortest path length of the real network, L_{random} is the shortest path length of the real network. when $\sigma > 1$, indicates that the network has a "small-world" topology.

Global Efficiency

The global efficiency E_{global} of the node is a measure which is similar to the shortest path length L of the node, but E_{global} completely avoids the problem of network disconnection. The formula is:

$$E_{global} = \frac{1}{N(N-1)} \sum_{i \neq j} \frac{1}{d_{ij}} \tag{6}$$

N represents the number of nodes, and d_{ij} represents the distance between node i and node j.

Local Efficiency

Relatively, like the clustering coefficient, local efficiency is also a parameter that describes the local information transmission capability of the network. The formula is:

$$E_{local} = \ <E_{local}(i)> \ = \frac{1}{N}\sum_i E_{local}(i) \qquad (7)$$

$$E_{local}(i) = \frac{1}{N_{G_i}(N_{G_i}-1)}\sum_{j\neq k} l_{j,k} \qquad (8)$$

Where G_i represents a subgraph composed of adjacent nodes of node i, and $l_{j,k}$ represents the least number of edges in the path of the two nodes.

3 Results

Small-World Topology
In this paper, we compared the changes of topological metrics on the global level for emotion self-regulation training. The two-sample t-test was conducted between the EG and the CG group, the comparison of the small-world network topology of the EG and the CG is as follows (see Fig. 2). the results showed that between 0.1–0.5 of the sparsity, gamma (which is represented by clustering coefficient, see Fig. 2(a)) of the EG and the CG group was significantly greater than 1, lambda (which is represented by shortest path length, see Fig. 2(b)) was greater than 1, and sigma (which is represented by small-world topology, see Fig. 2(c)) was also greater than 1, and the brain networks of both groups appeared small-world network properties.

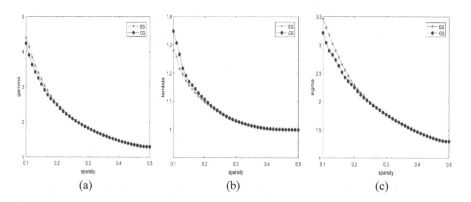

Fig. 2. Comparison of small world attributes between the EG and the CG

The paired t-test was conducted in the EG, the result of small-world metrics was the same as shown above. The result showed that all the EG presented small-world topology, including the EG after rt-fMRI-NF training (see Fig. 3). Therefore, for small-world attribute, we summarized that the participants in the EG and the CG group were significantly different in the small-world topology of the brain network in emotion self-regulation training.

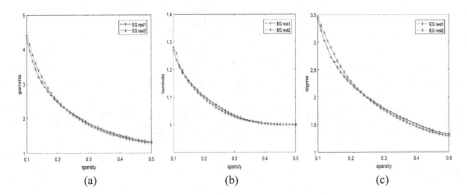

Fig. 3. Comparison of small world attributes in the EG after rt-fMRI emotion self-regulation training

Network Efficiency

The two-sample t-test was conducted between the EG and the CG, the results showed that, there was no significant difference between the global efficiency and the local efficiency of both the EG and the CG (see Fig. 4), when the sparsity is between 0.1–0.5.

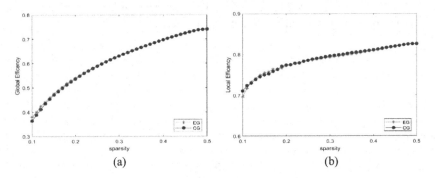

Fig. 4. Comparison of the global efficiency and the local efficiency between the EG and the CG

The paired t-test was conducted in the EG, there was no significant difference in the global network efficiency (see Fig. 5(a)). Meanwhile, when the sparsity is 0.1–0.2, the local efficiency has significantly improved (see Fig. 5(b)). This means that under these sparsity levels, the information transmission capacity of the brain network has increased. When the sparsity is 0.1–0.2 (when sparsity is between 0.12–0.14, $p < 0.05$), the density of brain network connections before and after neurofeedback is a good choice for the local efficiency.

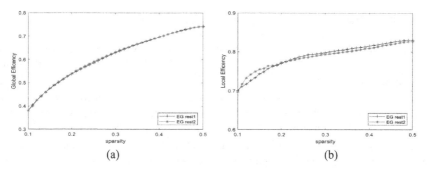

Fig. 5. Comparison of the global efficiency and the local efficiency in the EG after rt-fMRI emotion self-regulation training

4   Discussion

In this paper, we applied graph-based theoretical approaches to analyze the effect of emotion self-regulation training in certain parameters of the human brain network [14]. Our results show that in the neural feedback experiment, the human brain network presents a small-world topology, which is further validating the previous studies. When the sparsity is between 0.1–0.2, the small-world network attributes of the EG and the CG are different. Previous studies have shown that some mental illnesses (depression, traumatic stress disorder, autism, etc.) can lead to changes in the small-world attributes of the brain's functional network [8, 15]. Our results show that emotional self-regulation training can affect the small-world attributes of human brain networks. It means that the brain network trained by rt-fMRI neurofeedback maintains high efficiency in information transmission and processing, which may provide a new direction for the treatment of mental illness.

In terms of network efficiency attributes, the local network efficiency evaluates the local information transmission ability, and the results show that the local efficiency of the participants has increased, indicating that the brain function network has enhanced the local information processing capability. When the sparsity is between 0.1–0.2 (when sparsity is between 0.12–0.14, $p < 0.05$), the training is improved for local network efficiency. However, the global network efficiency has hardly changed in emotion self-regulation training. Global network efficiency assesses the overall efficiency of the brain function network. Experimental results show that the role of emotion self-regulation in overall network efficiency is very weak. This is explicable, because our emotional feedback only works on the brain area associated with emotions, so there is no significant change in overall network efficiency. Furthermore, our results are consistent with clinical findings, demonstrating the effectiveness of rt-fMRI neural feedback for brain network regulation.

5 Conclusion

Because of the complexity of the human brain, it is a very efficient way to study the brain as a network which consist of edges and nodes. In this paper, we used the graph theory approach to study the changes of some properties of human brain before and after neurofeedback experiments. Our results suggest that some of the brain's network topological properties change after emotion self-regulation training. We can further confirm that emotion self-regulation training is effective in enhancing the brain's ability to regulate the emotions of the brain, which may bring new ideas to the treatment of clinical diseases. For example, patients with depression and anxiety can improve emotional self-regulation ability by conducting scientific experimental training instead of medical treatment.

Acknowledgment. This work was supported by the National Key Research and Development Plan of China under grant 2017YFB1002502, the National Natural Science Foundation of China (No. 61701089), and the Natural Science Foundation of Henan Province of China (No. 162300410333).

References

1. Bassett, D.S., Bullmore, E., Verchinski, B.A., Mattay, V.S., Weinberger, D.R., Meyer-Lindenberg, A.: Hierarchical organization of human cortical networks in health and schizophrenia. J. Neurosci. Off. J. Soc. Neurosci. **28**(37), 9239–9248 (2008)
2. Bullmore, E., Sporns, O.: Complex brain networks: graph theoretical analysis of structural and functional systems. Nat. Rev. Neurosci. **10**(3), 186–198 (2009)
3. Watts, D.J., Strogatz, S.H.: Collective dynamics of 'small-world' networks. Nature **393**(6684), 440–442 (1998)
4. Colizza, V., Flammini, A., Serrano, M.A., Vespignani, A.: Detecting rich-club ordering in complex networks. Nat. Phys. **2**, 110–115 (2006)
5. Boccaletti, S., Latora, V., Moreno, Y., et al.: Complex networks: structure and dynamics. Phys. Rep. **2006**(424), 175–308 (2006)
6. Salvador, R., Suckling, J., Coleman, M.R., Pickard, J.D., Menon, D., Bullmore, E.: Neurophysiological architecture of functional magnetic resonance images of human brain. Cereb. Cortex **15**(9), 1332–1342 (2005)
7. van den Heuvel, M.P., Stam, C.J., Kahn, R.S., et al.: Efficiency of functional brain networks and intellectual performance. J. Neurosci. **29**, 7619–7624 (2009)
8. Ye, M., Yang, T., Peng, Q., Xu, L., Jiang, Q., Liu, G.: Changes of functional brain networks in major depressive disorder: a graph theoretical analysis of resting-state fMRI. PLoS ONE **10**(9), e0133775 (2015)
9. Zhang, J.R., et al.: Disrupted brain connectivity networks in drug-naive, first-episode major depressive disorder. Biol. Psychiatry **70**(4), 334–342 (2011)
10. Li, Z., et al.: Altered resting-state amygdala functional connectivity after real-time fMRI emotion self-regulation training. BioMed Res. Int. **2016** (2016). Article ID 2719895
11. Yan, C., Zang, Y.: DPARSF: a MATLAB toolbox for "pipeline" data analysis of resting-state fMRI. Front. Syst. Neurosci. **4**, 13 (2010)

12. Tzourio-Mazoyer, N., Landeau, B., Papathanassiou, D., et al.: Automated anatomical labeling of activations in SPM using a macroscopic anatomical parcellation of the MNI MRI single-subject brain. Neuroimage **15**(1), 273–289 (2002)

13. Wang, J., Wang, X., Xia, M.: Corrigendum: GRETNA: a graph theoretical network analysis toolbox for imaging connectomics. Front. Hum. Neurosci. **9**, 386 (2015)

14. Qin, Y., Tong, L., Gao, H., Wang, L., Zeng, Y., Yan, B.: Research on functional brain networks topological properties by real-time fMRI emotion self-regulation training. In: 2018 IEEE 3rd International Conference on Signal and Image Processing (ICSIP), Shenzhen, pp. 86–90 (2018)

15. Suo, X., et al.: Disrupted brain network topology in pediatric posttraumatic stress disorder: a resting-state fMRI study. Hum. Brain Mapp. **36**(9), 3677–3686 (2015)

Application of Convolutional Neural Network in Segmenting Brain Regions from MRI Data

Hafsa Moontari Ali[1], M. Shamim Kaiser[2](✉)(iD), and Mufti Mahmud[3](✉)(iD)

[1] Department of Computer Science and Engineering, Jahangirnagar University,
Savar 1342, Dhaka, Bangladesh
hafsa.moontari.ali@juniv.edu
[2] Institue of Information Technology, Jahangirnagar University,
Savar 1342, Dhaka, Bangladesh
mskaiser@juniv.edu
[3] Department of Computing and Technology, Nottingham Trent University,
Clifton, Nottingham NG11 8NS, UK
mufti.mahmud@gmail.com, mufti.mahmud@ntu.ac.uk

Abstract. Extracting knowledge from digital images largely depends on how well the mining algorithms can focus on specific regions of the image. In multimodality image analysis, especially in multi-layer diagnostic images, identification of regions of interest is pivotal and this is mostly done through image segmentation. Reliable medical image analysis for error-free diagnosis requires efficient and accurate image segmentation mechanisms. With the advent of advanced machine learning methods, such as deep learning (DL), in intelligent diagnostics, the requirement of efficient and accurate image segmentation becomes crucial. Targeting the beginners, this paper starts with an overview of Convolutional Neural Network, the most widely used DL technique and its application to segment brain regions from Magnetic Resonance Imaging. It then provides a quantitative analysis of the reviewed techniques as well as a rich discussion on their performance. Towards the end, few open challenges are identified and promising future works related to medical image segmentation using DL are indicated.

Keywords: Machine learning · Brain imaging · Neuroimaging · Segmentation · Deep learning · MRI

1 Introduction

Medical diagnosis, disease detection, treatment and response of treatment are key features and strategies in clinical medicine. Accurate diagnosis and early disease detection can assist doctors to ensure appropriate treatment for patients. Reliable analysis of medical images acquired from various imaging techniques (such as X-ray, Computed Tomography (CT), Magnetic Resonance Imaging (MRI),

© Springer Nature Switzerland AG 2019
P. Liang et al. (Eds.): BI 2019, LNAI 11976, pp. 136–146, 2019.
https://doi.org/10.1007/978-3-030-37078-7_14

Positron Emission Tomography (PET), Optical Coherence Tomography (OCT), Endoscopy) plays an important role in diagnostic medicine. Such analysis provides clear understanding of the disease localisation which then leads to appropriate treatment followed by monitoring of treatment response in patients.

Image segmentation – a (semi)-automatic detection of boundaries in an image – is a mandatory steps of image analysis pipeline. The high degree of variability in medical images make the segmentation laborious which assists to achieve further insights in the extracted image boundary.

This work presents a survey on Application of Convolutional Neural Network (CNN) in segmenting brain regions from MRI.

The organization of this paper is as follows: Sect. 2 describes the basic of CNN architecture, Sect. 3 discusses its application in medical image segmentation, Sect. 4 presents its performance evaluation and Sect. 5 lists some open challenges and probable research opportunities.

2   The Convolutional Neural Network

In general, deep learning (DL) architectures learn deep features from data in hierarchical and composite ways. This is achieved through multiple levels of abstraction where higher levels of abstraction are built on the top of lower levels and the lowest ones are the original input data [22] (Fig. 1).

This paper focuses only on CNN due to its wide acceptance as the most popular architecture for image analysis. CNN is a fully trainable biologically inspired version of multi-layer perceptron composed of alternating convolution and pooling layers followed by a fully connected layer at the end. Usually, CNN requires a large amount of data as the number of parameters and node needed to be trained is relatively high [22]. Some widely used CNN configurations include AlexNet, VGGNet and GoogLeNet. In CNN, an input image is convolved with kernels (k_j^l), biases (b_j^l) are added and a new feature map (x_j^l) is generated, such that, $x_j^l = f\left(\sum_{i \in M_j} x_i^{l-1} * k_{i}^l j + b_j^l\right)$. Unlike traditional ML approaches, CNNs learn and optimise best sets of convolution kernels. Incorporating such kernels with appropriate pooling operation extracts relevant features for a given task (e.g., classification, segmentation or recognition). CNNs are used in segmentation by classifying each pixel individually and presenting it with patches extracted around the particular pixel.

Fully Convolutional Network (FCN)– a variant of CNN– can generate a likelihood map instead of single pixel. FCN based segmentation network contains two paths: downsampling path (for semantic information) and upsamping path (for spatial information). In addition, skip connections are used to fully get the fine-grained spatial information [44].

U-Net is inspired from FCN which consists of contracting and expanding paths to capture the contextual and localisation information respectively. The segmentation map generated by U-Net contains only the pixels, so the full context is available in the input image [35].

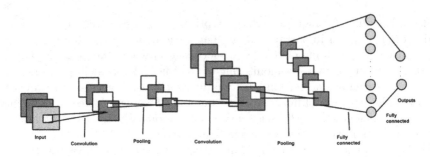

Fig. 1. Architecture of CNN

Dual pathway architecture is generated by coupling another identical convolutional pathway. for handling multi-class problem and processing multiple sources/multi-scale information from channels in input layers. The parallel pathways extract features independently from multiple channels [16,26].

In dilated CNN (Di-CNN), the weights of convolutional layers are sparsely distributed over a large receptive field. It is an effective architecture to achieve large receptive field with limited number of convolutional layers without subsampling layers [27,29].

3   CNN in Segmenting MRI

DL, in particular, CNN-based segmentation of MRI has brought significant change to clinical studies towards accurate, efficient and safe evaluation [23]. MRI–a noninvasive imaging–provides detailed functional and anatomical information of soft tissues, bones and organs in any arbitrary plane. The automatic segmentation of MRI is a key step in delineating the contour and any unusual anatomical structure or interior region.

A significant number of studies related to segmentation of MRI using CNN and its variant have been reported in the literature. Moeskops et al. introduced CNN based tissue regions segmentation from brain MRI. CNN combines multiple patches and different sized kernels to learn multi-scale features by estimating both intensity and spatial characteristics [25]. Brebisson et al. applied CNN for anatomical brain segmentation [3]. The authors have benchmarked it against multi-atlas methods of 2012 MICCAI challenge.

Moeskops et al. and Milletari et al. introduced adversarial training with DiCNN [27] and Hough CNN [24] for brain MRI segmentation. Moeskops et al. also employed single trained CNN for multiple segmentation tasks including a number of tissues in MR brain images, the pectoral muscle in MR breast images, and the coronary arteries in cardiac CTA, etc. [26].

The accurate segmentation of brain into white matter (WM), gray matter (GM) and cerebrospinal fluid (CSF) is an open challenges. Nguyen et al. used Gaussian mixture model (GMM) with CNN to segment brain MRI into WM, GM and CSF [30]. Moeskops et al. applied dilated CNN to segment isointense infant brain into WM, GM and CSF [29]. The dilated triplanar network is combined with an un-dilated 3D CNN. 3D FCN is also used for WM, GM and CSF segmentation in isointense infant brain MR [44]. Multi-scale deep supervision is applied to reduce gradient vanishing problem. Xu et al. combined three FCNs fusing with predicted probability map for small WM segmentation [42]. WM hyperintensities (WMH) are signal abnormalities in the white matter, and Ghafoorian et al. trained a CNN with non-uniformly sampled patches for WMH segmentation [9]. The authors also proposed a CNN-based approach in [10] for WMH segmentation combining the anatomical location information with network. WMH has some features similar to features of lesions and difficult to distinguish. Guerrero et al. developed a FCN architecture to simultaneously segment WMH and stroke lesions. The convolutional architecture used residual element to mitigate model complexity and improve optimization speed [11]. Moeskops et al. applied CNN to segment brain tissues and WMH at the same time [28]. Rachmadi et al. segmented WMH by combining the location information and global spatial information with CNN in the convolution level [34].

Choi et al. proposed a CNN-based approach for segmenting very small region such as striatum [5]. Instead of using the whole input image to segment such a small region, the authors employed two CNNs, known as local and global. The global CNN determined probable locations of striatum and the local CNN to predict the accurate labels of striatum voxels.

Shakeri et al. proposed a FCN based approach for the segmentation of frontostriatal sub-cortical structures in MR images [37]. The output of the segmentation approach was improved by feeding it as input to Markov Random Field (MRF). Kushibar et al. developed another sub-cortical brain structure segmentation approach by combining convolutional and spatial features [18]. It employed a 2.5D CNN model and the combination of spatial features was inspired from this work [10]. 3D FCN is also employed to segment sub-cortical brain region in MRI [8]. This method avoided over fitting by generating a deep network combining small convolution kernels. He et al. segmented left and right caudate nucleus along with left and right hippocampus from brain MRI using 3D FCN and 3D U-Net [13]. Here, affine invariant spatial information is learnt from FCN and used in 3D U-net. Another 3D CNN, called DeepNAT, was proposed in [40]. Bao et al. applied multi-scale structured CNN (MS-CNN) to segment several sub-cortical structures from MRI. They used dynamic random walker (DRW) as post-processing [1].

Pereira et al. proposed CNN-based method for segmentation of glioma from MRI. The authors stacked small convolution kernel to generate deeper CNN [33]. Hussain et al. applied CNN for glioma segmentation [15]. The architecture introduced two phase training method for data imbalance problem.

Glioblastoma is a highly malignant grade IV glioma. Yi et al. applied 3D CNN for glioblastoma segmentation. The first layer of the CNN was pre-defined with Difference-of-Gaussian (DoG) filters to learn 3D tumor MRI data [43].

Hoseini et al. [14] and Havaei et al. [12] applied CNN for brain tumor segmentation from MRI. Zhao et al. integrated FCN and conditional random field

(CRF) for tumor segmentation [46]. In [38], 3-FCNs are combined for tumor segmentation. Zhuge et al. applied holistically-nested network (HNN), an extension of CNN for high grade glioma (HGG) segmentation [48]. Deep supervision is performed in HNN through an additional weighted-fusion output layer (Table 1).

Table 1. Overview of papers using deep learning in MRI

Segmented region	Deep learning algorithm
Anatomical brain region	CNN [3]
Stratium	CNN [5]
Isointense infant brain	Di-CNN [29], CNN [25,45], FCN [17,31,32], 3D FCN [7,44]
WM, GM and CSF	CNN [30], 3D-FCN [6], FCN [42]
WMH	CNN [9,10,34]
WMH and stroke lesion	FCN [11]
WMH and brain tissue	CNN [28]
Sub-cortical brain structure	FCN [8,37],[1,18],3D {FCN,U-Net} [21], 3D CNN [40]
Brain tumor	CNN [12,14,33], FCN, CRF [38,46], 3D-CNN [43], OM-Net [47], H-CNN [48], DCNN [15], DRFCN, LS [20],
Brain lesion	3D CNN, CRF [16]
Brain	H-CNN [24], Di-CNN [27]
NPC	CNN [41]
Meningioma	DLM [19]
MS Lesion	CNN [2], CEN [4], C3CNN [39], FCN [36]

Kamnitsas et al. applied a dual pathway, 3D CNN with 3D fully connected CRF to segment brain/Ischemic stroke lesion [16]. Wang and Zu et al. applied CNN to segment Nasopharyngeal Cancer(NPC) and achieved performance similar to manual segmentation [41]. Laukamp et al. proposed a multiparametric deep-learning model (DLM) for detection and segmentation of meningiomas from MR images [19].

Le et al. combined recurrent FCN with level set (LS) method to develop a novel end-to-end approach (Deep Recurrent Level Set (DRLS)) [20]. Zhang et al. applied CNN to segment brain tissue in isointense stage [45]. The multiple intermediate layers of CNN are incorporated with multimodal brain information collected from T1/T2 and fractional anisotropy images as input feature maps. Isointense infant brain image segmentation is expanded and improved by Nie et al. using FCN [32]. The authors trained a separate network for each image and combined the outputs with higher network layers [32]. It optimizes the weights and biases for each modality corresponding to kernel size. FCN is also applied for infant brain segmentation in [17] and [31]. Dolz et al. proposed a densly connected 3D FCN in [7] for multimodal isointense brain segmentation. Dense connectivity exists between MR T1 and T2 in 3D context.

Multiple sclerosis (MS) is a chronic disease that causes scarred tissues (called lesions) being developed in brain/spinal cord. MRI scans show the symptoms of MS lesion as black holes. Brosch et al. proposed a novel approach for MS lesion segmentation [4] based on CEN and U-net. Valverde et al. [39] and Birenbaum et al. [2] proposed CNN based MS lesion segmentation. The first layer predicts the possible lesion areas and the second layer reduces the number of misclassified candidate pixels. Roy et al. applied FCN for MS lesion segmentation [36]. This is the first multi-view CNN segmentation approach in MRI that uses longitudinal information along with other features.

4   Performance Evaluation

Three well-known measures, the accuracy, dice coefficient, F1 score, were used to find the performance of a network architecture.

In brain tissue segmentation, the combination of FCN and DiCNN with adversarial training achieved superior dice score (DS) of 0.92 which outperforms other methods [27]. A multi-scale 3-layer CNN [25] outperformed multivariate MRF method (0.827 vs. 0.737). A multi-scale 4-layer CNN with weight sharing and location information achieved competitive DS of 0.791 [10]. Consecutive application of CNN-DNN with GMM achieved better performance than kNN (0.86 vs. 0.73) [30]. A CNN trained with non-uniform sampled patches outperformed a similar CNN with uniformly sampled patches (0.78 vs. 0.73) [9]. The uResNet architecture, containing 8-residual layers, 3-deconvolutional layers and a convolutional layer performed better than Lesion prediction algorithm (0.695 vs. 0.647) [11]. A 9-layer CNN architecture acquired DS of 0.67 [28], whereas another CNN using GSI achieved DS of 0.54 [34]. However, the latter one outperformed DBM which achieved only 0.33. The performance of FCN architectures were analysed in [42], where 3-FCNs outperformed 1-FCN (0.78 vs. 0.7). In segmenting striatum, two CNNs, local and global CNN outperformed FreeSurfer on OASIS dataset (0.893 ± 0.017 vs. 0.786 ± 0.015) and achieved competitive score on IBSR dataset (0.826 ± 0.038 vs. 0.827 ± 0.022) [5]. A 5-layer FCN and CRF was applied on IBSR dataset to segment the 3D T1-weighted MR images into thalamus, caudate, putamen and pallidum (DS: Proposed: 0.87 vs Bayesian: 0.85) [37].

The 3D FCN outperformed 2D FCN by achieving 0.92 validated on both ABIDE and IBSR dataset [8]. While segmenting caudate nucleus and hippocampus, FCN guided with shape learning network outperformed 3D U-Net on LPBA40 dataset (0.80175 vs. 0.779) [13]. CNN based methods reported outperformed the atlas-based methods (CNN:0.869 vs. LF-MA:0.867; 3D CNN: 0.906 vs. MA:0.904) on MICCAI 2012 and IBSR 18 datasets. MS-CNN based approach outperformed majority voting (MV) and patch-based label fusion (PBL) on IBSR dataset (MS-CNN: 0.807, MV: 0.658, PBL: 0.760) and LPBA40 dataset

(MS-CNN: 0.843, MV: 0.764, PBL: 0.843). The performance improved on both datasets when MS-CNN was followed by DRW (MS-CNN+DRW: 0.822 vs MS-CNN: 0.807 on IBSR dataset), (MS-CNN+DRW: 0.850 vs MS-CNN: 0.843 on LPBA40 dataset) [1] (Fig. 2).

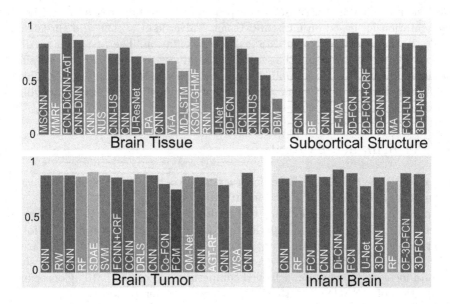

Fig. 2. Performance comparison of Various CNN based DL techniques when applied to Brain MRI in: tissue, subcortical structure, tumor and infant brain segmentation.

To segment HGG and LGG, CNN based methods achieved similar performances comparing with other methods (CNN: 0.88 vs. RW:0.88, TWOPATHCNN: 0.88 vs. RF: 0.87) both validated on BRATS 2013 challenge dataset [33], [12]. FCN, trained with image patches and CRF with image slices have outperformed cascaded CNN evaluated in BRATS 2013, BRATS 2015 and BRATS 2016 datasets (0.86 vs. 0.84) [46]. HNN outperformed FCN (0.78 vs. 0.61) in HGG tumor segmentation evaluated on 20 data from BRATS 2013 training dataset [48]. Two phase weighted trained CNN model performed better than anatomy guided RF model (0.86 vs. 0.85) on both BRATS 2013 and 2015 datasets [15]. A 5-layer 3D CNN based approach outperformed other methods to segment whole, core and active tumor on BRATS 2015 dataset (3D CNN: 0.89, Appearance and Context Sensitive Features: 0.83, Extremely randomized tree: 0.83) [43]. CNN has significantly outperformed Watershed algorithm for tumor segmentation on 15 T1-weighted MR images of NPC patients (0.79 vs. 0.6) [41]. In another study, CNN with task-level parallelism outperformed other methods (CNN: 0.90, LSE-KMC: 0.84). OM-Net [47], a CNN architecture achieved DS of 0.87 beating U-Net shape inspired architecture (DS: 0.85). But CNN was unable to score better result in [20] in comparison with DRLS (DRLS: 0.89,

CNN: 0.88). Three concurrent FCNs, each to train 3 different filtered (Gaussian, Gabor and median) multi-MR images are applied for tumor segmentation and achieved superior result than Fuzzy C-Means (FCM) (0.8 vs. 0.75) [38]. The performance of a 11-layer DeepMedic architecture with CRF is outperformed over RF and an ensemble of three networks to segment Traumatic brain injuries (RF: 0.511 ± 0.20, RF+CRF: 0.548 ± 0.185, DeepMedic: 0.623 ± 0.164, DeepMedic+CRF: 0.63 ± 0.163), Brain tumor (Ensemble+CRF: 0.849, Ensemble: 0.845, DeepMedic+CRF: 0.847, DeepMedic: 0.836) Ischemic Stroke lesion (DeepMedic: 0.64 ± 0.23, DeepMedic+CRF: 0.66 ± 0.24) segmentation [16]. To segment infant brain images into WM, GM and CSF, CNN has outperformed RF by combining multi-modality T1, T2 and FA images (0.8503 ± 0.0227 vs. 0.8351 ± 0.252) [45]. In a similar segmentation task, multi-modality FCN also outperformed RF for WM (0.887 ± 0.021 vs. 0.841 ± 0.027) [31]. FCN outperformed U-Net (0.889 vs 0.796) [17]. For more accuracy, 3D FCN with convolution and concatenate (CC) is applied on T1 and T2 images [31] and it obtained better dice value than RF (0.94 ± 0.0075 vs. 0.8765 ± 0.0112). Another 3D FCN with context information is able to segment WM, GM and CSF from iSeg-2017 dataset superior to 3D FCN without context information (0.922 ± 0.008 vs. 0.916 ± 0.008) [44]. A 7-layer dilated triplanar CNN and 12-layer non-dilated 3D CNN are evaluated in MICCAI grand challenge on 6-month infant brain MRI segmentation into WM, GM and CSF with dice value 0.894.

5 Limitations and Challenges

Deep learning framework can learn intense features from huge imaging dataset. However, some challenges and future perspectives are discussed below:

Creating label data requires huge processing which is an open challenge for designing supervised architecture for image analysis. However the accuracy can be improved considerably using semantic segmentation which can also be explored. Inter-organizational collaborations are essential for generating gigantic volume of label data to eliminate the resource limitation problem.

Training a classifier with a large volume of collected data can biased towards majority classes and will provide low accuracy.

To handle real time segmentation of big imaging data, distributed as well as parallel computing infrastructure are required.

6 Conclusion

Automatic segmentation of medical images plays substantial role in computer-aided medical image analysis pipeline. This paper presents the use of CNN in image segmentation for diagnosis, disease detection, treatment and response of treatment from MRI. The performance of different CNN architectures has been evaluated for MRI taken from various brain tissues and/or regions. The results show that the CNN and its variant based architectures are popular in medical image segmentation. At the end, open challenges have been identified and research possibilities are outlined which can be utilized to improve the performance of DL based medical image analysis.

References

1. Bao, S., Chung, A.C.: Multi-scale structured CNN with label consistency for brain MRI segmentation. Comput. Methods Biomech. Biomed. Eng. Imaging Vis. **6**(1), 113–117 (2018)
2. Birenbaum, A., Greenspan, H.: Longitudinal multiple sclerosis lesion segmentation using multi-view convolutional neural networks. In: Carneiro, G., et al. (eds.) LABELS/DLMIA - 2016. LNCS, vol. 10008, pp. 58–67. Springer, Cham (2016). https://doi.org/10.1007/978-3-319-46976-8_7
3. de Brebisson, A., Montana, G.: Deep neural networks for anatomical brain segmentation. In: Proceedings of IEEE CVPR, pp. 20–28 (2015)
4. Brosch, T., et al.: Deep 3D convolutional encoder networks with shortcuts for multiscale feature integration applied to multiple sclerosis lesion segmentation. IEEE Trans. Med. Imaging **35**(5), 1229–1239 (2016)
5. Choi, H., Jin, K.H.: Fast and robust segmentation of the striatum using deep convolutional neural networks. J. Neurosci. Methods **274**, 146–153 (2016)
6. Dolz, J., et al.: HyperDense-Net: a hyper-densely connected CNN for multi-modal image segmentation. IEEE Trans. Med. Imaging **38**(5), 1116–1126 (2019)
7. Dolz, J., Ayed, I.B., Yuan, J., Desrosiers, C.: Isointense infant brain segmentation with a hyper-dense connected CNN. In: Proceedings of ISBI 2018, pp. 616–620 (2018)
8. Dolz, J., Desrosiers, C., Ayed, I.B.: 3D fully convolutional networks for subcortical segmentation in MRI: a large-scale study. NeuroImage **170**, 456–470 (2017)
9. Ghafoorian, M., et al.: Non-uniform patch sampling with deep CNNs for white matter hyperintensity segmentation. In: Proceedings of ISBI, pp. 1414–1417 (2016)
10. Ghafoorian, M., et al.: Location sensitive deep CNNs for segmentation of white matter hyperintensities. Sci. Rep. **7**, 5110 (2017)
11. Guerrero, R., et al.: White matter hyperintensity and stroke lesion segmentation and differentiation using CNNs. NeuroImage: Clin. **17**, 918–934 (2018)
12. Havaei, M., et al.: Brain tumor segmentation with deep neural networks. Med. Image Anal. **35**, 18–31 (2017)
13. He, Z., Bao, S., Chung, A.: 3D deep affine-invariant shape learning for brain MR image segmentation. In: Stoyanov, D., et al. (eds.) DLMIA/ML-CDS - 2018. LNCS, vol. 11045, pp. 56–64. Springer, Cham (2018). https://doi.org/10.1007/978-3-030-00889-5_7
14. Hoseini, F., Shahbahrami, A., Bayat, P.: An efficient implementation of deep convolutional neural networks for MRI segmentation. J. Digit. Imaging **31**, 1–10 (2018)
15. Hussain, S., Anwar, S.M., Majid, M.: Segmentation of glioma tumors in brain using deep convolutional neural network. Neurocomputing **282**, 248–261 (2018)
16. Kamnitsas, K., et al.: Efficient multi-scale 3D CNN with fully connected CRF for accurate brain lesion segmentation. Med. Image Anal. **36**, 61–78 (2017)
17. Kumar, S., et al.: InfiNet: fully convolutional networks for infant brain MRI segmentation. In: Proceedings of ISBI 2018, pp. 145–148 (2018)
18. Kushibar, K., et al.: Automated subcortical brain structure segmentation combining spatial and deep convolutional features. Med. Image Anal. **48**, 177–186 (2018)

19. Laukamp, K.R., et al.: Fully automated detection and segmentation of meningiomas using deep learning on routine multiparametric MRI. Eur. Radiol. **29**(1), 124–132 (2018)

20. Le, T.H.N., Gummadi, R., Savvides, M.: Deep recurrent level set for segmenting brain tumors. In: Frangi, A.F., Schnabel, J.A., Davatzikos, C., Alberola-López, C., Fichtinger, G. (eds.) MICCAI 2018. LNCS, vol. 11072, pp. 646–653. Springer, Cham (2018). https://doi.org/10.1007/978-3-030-00931-1_74

21. Li, X., et al.: 3D multi-scale FCN with random modality voxel dropout learning for intervertebral disc localization and segmentation from multi-modality MRI. Med. Image Anal. **45**, 41–54 (2018)

22. Mahmud, M., Kaiser, M.S., Hussain, A., Vassanelli, S.: Applications of deep learning and reinforcement learning to biological data. IEEE Trans. Neural Netw. Learn Syst. **29**(6), 2063–2079 (2018)

23. Miller, C.G., Krasnow, J., Schwartz, L.H.: Medical Imaging in Clinical Trials. Springer, London (2014). https://doi.org/10.1007/978-1-84882-710-3

24. Milletari, F., et al.: Hough-CNN: deep learning for segmentation of deep brain regions in MRI and ultrasound. Comput. Vis. Image Underst. **164**, 92–102 (2017)

25. Moeskops, P., et al.: Automatic segmentation of mr brain images with a CNN. IEEE Trans. Med. Imaging **35**(5), 1252–1261 (2016)

26. Moeskops, P., et al.: Deep learning for multi-task medical image segmentation in multiple modalities. In: Ourselin, S., Joskowicz, L., Sabuncu, M.R., Unal, G., Wells, W. (eds.) MICCAI 2016. LNCS, vol. 9901, pp. 478–486. Springer, Cham (2016). https://doi.org/10.1007/978-3-319-46723-8_55

27. Moeskops, P., Veta, M., Lafarge, M.W., Eppenhof, K.A.J., Pluim, J.P.W.: Adversarial training and dilated convolutions for brain MRI segmentation. In: Cardoso, M.J., et al. (eds.) DLMIA/ML-CDS - 2017. LNCS, vol. 10553, pp. 56–64. Springer, Cham (2017). https://doi.org/10.1007/978-3-319-67558-9_7

28. Moeskops, P., et al.: Evaluation of a deep learning approach for the segmentation of brain tissues and WMH of presumed vascular origin in MRI. NeuroImage: Clin. **17**, 251–262 (2018)

29. Moeskops, P., Pluim, J.P.W.: Isointense infant brain MRI segmentation with a dilated convolutional neural network. CoRR abs/1708.02757 (2017)

30. Nguyen, D.M., et al.: 3D-brain segmentation using deep neural network and Gaussian mixture model. In: Proceedings of WACV 2017, pp. 815–824 (2017)

31. Nie, D., et al.: 3-D fully convolutional networks for multimodal isointense infant brain image segmentation. IEEE Trans. Cybern. **49**(3), 1123–1136 (2019)

32. Nie, D., Wang, L., Gao, Y., Sken, D.: FCNs for multi-modality isointense infant brain image segmentation. Proceedings of ISBI 2016, pp. 1342–1345 (2016)

33. Pereira, S., et al.: Brain tumor segmentation using CNNs in MRI images. IEEE Trans. Med. Imaging **35**(5), 1240–1251 (2016)

34. Rachmadi, M., et al.: Segmentation of WMHs using CNNs with global spatial information in routine clinical brain MRI with none or mild vascular pathology. Comput. Med. Imaging Graph. **66**, 28–43 (2018)

35. Ronneberger, O., Fischer, P., Brox, T.: U-Net: convolutional networks for biomedical image segmentation. In: Navab, N., Hornegger, J., Wells, W.M., Frangi, A.F. (eds.) MICCAI 2015. LNCS, vol. 9351, pp. 234–241. Springer, Cham (2015). https://doi.org/10.1007/978-3-319-24574-4_28

36. Roy, S., et al.: Multiple sclerosis lesion segmentation from brain MRI via fully convolutional neural networks. arXiv preprint arXiv:1803.09172 (2018)

37. Shakeri, M., et al.: Sub-cortical brain structure segmentation using F-CNN's. In: Proceedings of ISBI 2016, pp. 269–272. IEEE (2016)

38. Shen, G., et al.: Brain tumor segmentation using concurrent fully convolutional networks and conditional random fields. In: Proceedings of ICMIP 2018, pp. 24–30 (2018)
39. Valverde, S., et al.: Improving automated multiple sclerosis lesion segmentation with a cascaded 3D CNN approach. NeuroImage **155**, 159–168 (2017)
40. Wachinger, C., Reuter, M., Klein, T.: DeepNAT: deep CNN for segmenting neuroanatomy. NeuroImage **170**, 434–445 (2018)
41. Wang, Y., et al.: Automatic tumor segmentation with deep convolutional neural networks for radiotherapy applications. Neural Process. Lett. **48**, 1–12 (2018)
42. Xu, B., et al.: Orchestral fully convolutional networks for small lesion segmentation in brain MRI. In: Proceedings of ISBI 2018, pp. 889–892. IEEE (2018)
43. Yi, D., Zhou, M., Chen, Z., Gevaert, O.: 3-D convolutional neural networks for glioblastoma segmentation. CoRR abs/1611.04534 (2016)
44. Zeng, G., Zheng, G.: Multi-stream 3D FCN with MSDS for multi-modality isointense infant brain MRI segmentation. In: Proceedings of ISBI 2018, pp. 136–140 (2018)
45. Zhang, W., et al.: Deep convolutional neural networks for multi-modality isointense infant brain image segmentation. NeuroImage **108**, 214–224 (2015)
46. Zhao, X., et al.: A deep learning model integrating FCNNs and CRFs for brain tumor segmentation. Med. Image Anal. **43**, 98–111 (2018)
47. Zhou, C., Ding, C., Lu, Z., Wang, X., Tao, D.: One-pass multi-task convolutional neural networks for efficient brain tumor segmentation. In: Frangi, A.F., Schnabel, J.A., Davatzikos, C., Alberola-López, C., Fichtinger, G. (eds.) MICCAI 2018. LNCS, vol. 11072, pp. 637–645. Springer, Cham (2018). https://doi.org/10.1007/978-3-030-00931-1_73
48. Zhuge, Y., et al.: Brain tumor segmentation using holistically-nested neural networks in MRI images. Med. Phys. **44**(10), 5234–5243 (2017)

Informatics Paradigms for Brain and Mental Health Research

Time Recognition of Chinese Electronic Medical Record of Depression Based on Conditional Random Field

Shaofu Lin[1,2], Yuanyuan Zhao[2], and Zhisheng Huang[3,4(✉)]

[1] Beijing Institute of Smart City,
Beijing University of Technology, Beijing 100124, China
linshaofu@bjut.edu.cn
[2] Faculty of Information, Beijing University of Technology,
Beijing 100124, China
yuans@emails.bjut.edu.cn
[3] Advanced Innovation Center for Human Brain Protection,
Capital Medical University, Beijing, China
huang@cs.vu.nl
[4] Department of Computer Science, VU University Amsterdam,
1081 HV Amsterdam, The Netherlands

Abstract. As an important entity in medical texts, time information plays an important role in structuring medical information and supporting clinical decision-making. In this paper, time expressions in Chinese electronic medical record text of depression are studied. The method combines regular expressions with Conditional random fields (CRFs) to recognize time expressions in Chinese electronic medical records. The test data are realistic electronic medical records of depression provided by a hospital in Beijing. The proposed method uses regular expressions to initially recognize the explicit time expression in the text, and adds a dictionary of common drugs and symptoms of depression to the word segmentation, which increases the accuracy of word segmentation. External dictionary features are optimized, and dictionaries are divided into time modifier dictionary, time representation dictionary and event dictionary, which effectively improve the accuracy and recall rate of conditional random field recognition results. Experiments show that the accuracy and recall rate of this method are 96.75% and 93.33% respectively.

Keywords: Electronic medical record for depression · Time representation regular expression · Conditional random field · Named entity recognition

1 Introduction

Depression is a common mental disorder, which mainly manifests as depression, low interest, pessimism, slow thinking, lack of initiative and other symptoms. People with serious symptoms may have suicidal thoughts and behavior. At present, it has become

Supported by the Beijing High-level Foreign Talents Subsidy Program 2019 (Z201919).

P. Liang et al. (Eds.): BI 2019, LNAI 11976, pp. 149–158, 2019.
https://doi.org/10.1007/978-3-030-37078-7_15

the second most serious disease in the global disease which causes serious burden to human beings. According to official media data, the number of patients suffering depression in China has reached 90 million. About 1 million people commit suicide every year because of depression. However, only 5% of the patients suffering depression have received treatment. A large number of patients have not received timely diagnosis and treatment, which result in more serious consequences such as suicides. With the development of medical informationization, electronic medical records are applied in major hospitals. How to extract effective medical information from electronic medical record text has become one of the important issues related to national health and medical development. Among them, time information, as a common entity in medical texts, can reflect the chronological nature of patients' treatment and disease development. Many entities representing events are related to time [1]. This paper mainly studies the time recognition of Chinese electronic medical records of depression. It is of great significance for structuring electronic medical records, helping doctors to quickly understand the patient's situation and putting forward treatment plans. At present, there are two main ways to recognize time expression: one is a rule-based method, the other is a statistical machine learning method. Because of the diversity of time expression in medical record text expression, there will be incomplete template coverage when a rule-based method is used to recognize time expression. Under this circumstance, this paper uses a mixed method of combining rule and statistical machine learning to study time recognition in Chinese electronic medical record text using conditional random field (CRF) model.

2 Related Work

Time entities are often studied as part of named entity extraction. At the MUC conference held in 1998, the requirement of time evaluation was added to the task of named entity recognition for the first time. TempEval-2010 [2] proposed the task of time recognition, while expanding the language from English to Chinese, English, Italian, French, Korean, Spanish and other six languages. Clinical TempEval-2015 [3] extends time information recognition to the medical field, including time recognition, medical time recognition and time relationship recognition for 600 clinical notes and medical records from cancer patients.

At present, there is a certain research foundation for Chinese time expression recognition, but it mainly focuses on the fields of news, finance and so on. [4, 5] Jutha et al. [6] proposed a method of identifying time phrases based on CRF, which transformed the recognition problem into a sequence annotation problem. Wu et al. [7] used CRF to identify time units, and added event trigger lexicon and time affix lexicon to locate time expressions. The problem of inaccurate location of time expression is solved. Liu et al. [8] used CRF to recognize time expression. While extracting traditional features, they added semantic role features to construct feature vectors according to the characteristics of Chinese time expression to improve the recognition accuracy. At present, there are few studies on time expression recognition in Chinese electronic medical record text. Jindal et al. [9] used a rule-based method to identify the time expression in the medical records, and divided the time into seven patterns. Time expression is recognized by using Heidel-Time, a time recognition tool, combined with the combination rules of time in clinical narrative. The method has a good effect on time recognition of date type, and the errors mainly focus on the recognition of lesser known time expression and complex time expression. Zhou et al. [10] recognized the time expression in the medical record text based on regular expression, but because of the diversity of time expression in the Chinese medical record text, it is difficult to sort out all the regular expressions. So there will be the problem of identifying omissions. Zhu et al. [11] used CRF to recognize the time expression in Chinese online medical texts. Context markers and part-of-speech (POS) markers were used as basic features. A Dictionary of time words and numerals were constructed as external features. However, when a word is used as the processing unit to generate a set of feature vectors, wrong word segmentation will lead to error layer by layer transmission amplification, which will affect the accuracy of recognition.

It can be seen from the above that the CRF model is mostly used to realize the recognition of time expression at present, and the accuracy of time expression recognition can be improved by analyzing text corpora and adding corresponding features. The difficulties of time information recognition in Chinese electronic medical record mainly lie in the diversity of time expression, the accuracy of word segmentation and the inaccuracy of locating boundary in part of time expression. Aiming at the diversification of time expression in Chinese electronic medical record text, this paper divides time into clear time expression, fuzzy time expression and event-based time expression; adds a dictionary of common medicines and common symptoms of depression to the accuracy of word segmentation; adds external features to the long time expression. The method of combination of regular expression and time expression is used to recognize time expression.

The rest of this paper is organized as follows. Section 2 gives an overview of related works, Sect. 3 introduces the classification of time expression in Chinese electronic medical records of depression. Section 4 elaborates the recognition scheme of time expression in Chinese electronic medical records. Section 5 illustrates the experimental of time expression recognition and the analysis of experimental results. The last section includes conclusion and future work.

3 Classification of Time Expressions in Chinese Electronic Medical Record Texts

The Fig. 1 is an excerpt of the description of the current history of a depressive patient. It can be seen from the picture that there are nine time expressions in this brief description, which are closely related to the symptoms and treatment of the patient. It is of great significance to give specific treatment plans for the patient. And the expression format of time is different, not only different time has different expression format, but also the same time has different expression format. For example, ten days can be expressed as either a number of 10 days or a capitalized number of 10 days. For the time expression in the medical record, we can see that there are still many different time expressions, such as 20 min later, early March, after the application of dexamethasone. So this paper studies the text of Chinese electronic medical record of depression, and the expressions of time are divided into simple time and compound time. Based on TIMEX2 annotation specification [12], simple time expression is divided into clear time expression, fuzzy time expression and event-based time expression. Clear time expression refers to the specific time unit determined by the time expression. Fuzzy time expression refers to the time expression with fuzzy words such as approximate, surplus, left and right. Event-based time expression refers to the time expression of events such as discharge, treatment, surgery and so on. Compound time expression is the combination of several simple time expressions. The example is shown in the Table 1.

Table 1. Time categorization table of medical record text.

Type	Explain	Example
Date	The time expressed is one day or longer than one day, and the smallest granularity is one day, which can be defaulted	31 September 2018
Time point	The smallest granularity is in seconds, representing the time of the day. Combining time, seconds, points and numbers	9:30
Time period	A modifier that expresses a clear time span but does not indicate ambiguity	3 months, March to May 2018
Age	Use age to indicate time	Seven years old
Section time	A combination of time periods and fuzzy words denotes time points	Two weeks ago, four days later
Relative time	Without a clear number, you need to refer to the context information to determine the time point. The smallest granularity is day and the largest granularity is year	The day before yesterday, last weekend, next month
Dressing time	Composed of time modifiers plus time points or periods	More than a month, around 2009.
Time words	Words that can express time	The first ten days, the Mid-Autumn Festival
Event-based time expression	The name and verb of an event can be followed by a preposition or a localizer, and a time phrase can be represented by a combination of time information	One month after hospitalization, three days after appendicitis surgery
Compound time	Combination of Simple Time Expressions	One week after discharge in March 2018

At the beginning of March 2015, the patient again presented with symptoms of bad mood, upset, panic, chest tightness, tension, sweating, less speech and loss of appetite. On March 19, our hospital went to the clinic and took venlafaxine 150 mg/day for 10 days. The symptoms did not improve. On March 28, due to pulmonary inflammation, cefoperazone and sulbactam were intravenously infused to allergic reactions, which were alleviated by dexamethasone. Two hours later, the patient suddenly developed chest tightness, dyspnea, legs weakness, sweating, and glassy eyes. No abnormalities were found in the admission test. After 20 minutes, the patient recovered spontaneously. On the second day, the above symptoms appeared again, and the symptoms were relieved after half an hour.

Fig. 1. Example of case history description

4 Recognition Scheme

After analyzing the time expression in the text of Chinese medical record of depression, it is found that the fuzzy time type is mostly formed by time modifier and clear time expression. The event-based time expression is mostly composed of event and fuzzy time expression. Because the clear time expression is more regular, this paper uses regular expression to identify the clear time expression. Then CRF is used to recognize all time expressions in medical record text based on clear time expressions. The time recognition flow chart is shown in Fig. 2.

4.1 Recognition of Clear Time Expressions

Regular expressions are patterns used to match combinations of characters in strings. For the time recognition of rules, this method has a good effect.

In this paper, some regular expressions designed for several different types of explicit time expressions are shown in Table 2.

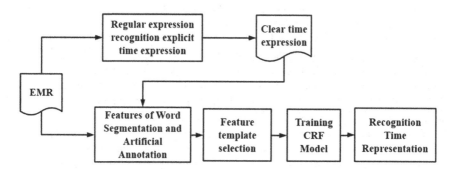

Fig. 2. Time recognition flow chart.

Table 2. Regular expression of partial time type

Time category	Regular expression
Date	`(((([12][0-9]{3})(([|/|\-|\_|\.]((([0]?[1-9])|([1][012]))(?=\D)))(([month|/|\-|\_|\.][0-3]?[0-9](day|numbers)?)|month)?)|year))`
Time point	`(((([01]?[0-9])|([2][0-4]))(([hours|point]((([0-5][0-9])|([6][0]))[minutes]((([0-5][0-9])|([6][0]))[seconds])?)|[o'clock|point]))`
Time period	`(?<![a-zA-Z0-9*])(((([12][0-9]{3})(([year|/|\-|_|\.]((([0]?[1-9])|([1][012]))(?=\D)))(([month|/|\-|_|\.][0-3]?[0-9](day|numbers)?)|month)?)|year))[to]|((((([ten]?[one,two,three,four,five,six,seven,eight,nine])|((([0]?[1-9])|([1][0-2])))(([month|/|\-|_|\.](([12][0-9])|([3][01])|([0]?[1-9])))(day|numbers)?)|month)(?![a-zA-Z0-9\%]))|(?<![a-zA-Z0-9*])((([12][0-9]{3})(([year|/|\-|_|\.]((([0]?[1-9])|([1][012]))(?=\D)))(([month|/|\-|_|\.][0-3]?[0-9](day|numbers)?)|month)?)|year)))`
Age	`((([0-9][.]?[0-9])|([half,one,two,three,four,five,six,seven,eight,nine,ten]+))[years old]`

4.2   CRF

CRF was proposed by Lafferty et al. in 2001, [13] is an undirected graph model, which combines the characteristics of maximum entropy model and hidden Markov model. It is a conditional probability distribution of another group of output random variables given a set of input random variables. In recent years, it has achieved good results in word segmentation, part-of-speech tagging and named entity recognition. CRF is a typical discriminant model. Its joint probability can be written in the form of several potential functions. The most commonly used one is linear chain conditional random field, which can effectively overcome the limitation of the assumptions of hidden Markov model and the problem of maximum entropy model labeling paranoia. Therefore, the linear chain conditional random field is also used in this paper.

Let x = (x1, x2,... Xn) denotes the sequence of input data observed, y = (y1, y2,... Yn) denotes a state sequence. Given an input sequence, the CRF model of a linear chain defines the joint conditional probability of the state sequence as follows:

$$P(Y|X) = \frac{1}{Z(X)} \exp\left(\sum_{i=1}^{n} \sum_{j} \lambda_j f_j(y_{i-1}, y_i, X, i)\right)$$

$$Z(X) = \sum_{j} \exp\left(\sum_{i=1}^{n} \sum_{j} f_j(y_{i-1}, y_i, X, i)\right)$$

$Z(X)$ is a probability normalization factor based on observation sequence X; $f_j(y_{i-1}, y_i, X, i)$ is an arbitrary eigenfunction, usually a binary function of 0 or 1; and λ_j is the weight of each eigenfunction.

4.3   Time Information Extraction Based on CRF

In this paper, the problem of time recognition is transformed into the problem of sequential tagging, and CRF is used for entity recognition.

The time information extraction model based on CRF is shown in Fig. 3.

Word Segmentation and POS Tagging. In this paper, we use hanLP tool [14] for word segmentation and part-of-speech tagging. When segmenting medical record text, we add a dictionary of commonly used drugs for depression and a dictionary of common symptoms to increase the accuracy of word segmentation. After word segmentation, the corpus is labeled artificially, and the time entity in the text is labeled by BIO marker. Each word type in the sentence is one of B, I and O markers. If a time expression consists of several words, then B denotes the first time word, the second time word is denoted by I, not all time words are denoted by O. For example, "Before and after the Mid-Autumn Festival in 2013, there were no obvious inducements of headache, dizziness, anxiety, insomnia and other symptoms" labeled as "before and after/B the Mid-Autumn Festival/I in 2013/I,/O without obvious/O inducements/O occurrence/O headache/O,/O dizziness/O,/O anxiety/O,/O insomnia/O and other/O symptoms/O".

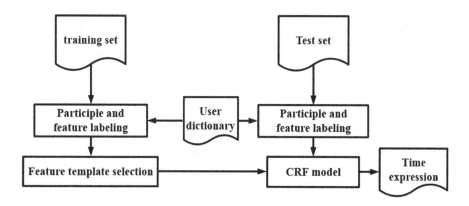

Fig. 3. Time information recognition model based on CRF

Feature Selection. This paper chooses POS, context and location features as the benchmark features. According to the characteristics of electronic medical record text, four features are added to identify the time expression, as shown in Table 3.

Table 3. Feature for time recognition

Feature type	Description
Part of Speech Features	Part of Speech of Current Words
Contextual Features	Pre-and Post-word Features of Current Words
Location Features	Location in Current Word Clauses
Time Representation Word Features	Is the Current Word Time Representation Word?
Event dictionary features	Whether the current word is an event of hospitalization, discharge, treatment, etc.
Characteristics of Time Modified Lexicon	Whether the current word is a temporal modifier, such as front and back, left and right, approximation, remainder, etc.
The characteristics of time numerals	Whether the current word contains zero, one, two or one, two, etc.

5   Experiments and Analysis

5.1   Corpus and Tools

In this study, 80 authentic medical records from depressive patients were selected randomly as training corpus, and the remaining 20 were used as test corpus. CRF uses CRF++ 0.58 and Perl script conlleval.Pl as evaluation tool.

5.2   Evaluation Index

The experimental results were evaluated by the calculated accuracy (P), recall (R) and F-measure (F1) values.

1. P = Number of time phrases correctly identified by the system/Number of time phrases correctly identified by the system*100%
2. R = Number of time phrases correctly recognized by the system/Number of time phrases in text*100%
3. F1 = 2PR/(P + R)*100%

5.3   Experimental Results and Analysis

The patient's electronic medical record includes current medical history, past medical history, personal history, family history and so on. There are a lot of redundant information in time information extraction research. Because the part of current medical history has a lot of time expression, this paper chooses the part of current medical history of 80 real medical records to study, and carries out experiments under different feature templates. The results are shown in Table 4.

Table 4. Experimental results under different feature templates

Feature template	P	R	F
Baseline Features	94.78%	82.58%	88.26%
Baseline + Numerals	91.15%	78.03%	84.08%
Baseline + Time Modifier	94.83%	83.33%	88.71%
Baseline + Time Modifier + Time Representatives	95.16%	89.39%	92.19%
Baseline + Time Modifier + Time Representatives + Event Dictionary	96.75%	90.15%	93.33%

From the above experimental data, it can be seen that by adding symptoms dictionary, drug dictionary and definite time expression dictionary identified by regular expressions in the previous segmentation, the recognition effect based on the baseline features achieves 94.78% accuracy, 82.58% recall rate and 88.26% F value. However, after numeral features are added, P, R and F values decrease slightly. This decline in P, R and F values is due to the sparsity of numeral features in medical records. The recognition effect is improved and the recall rate is improved obviously after adding the time representation feature. The results show that when the feature template is a baseline feature plus time modifier feature, time representatives feature and event dictionary feature, the effect of time recognition is the best, the accuracy rate is 96.75%, the recall rate is 90.15%, and the F value is 93.33%.

6   Conclusion

In this paper, time in Chinese electronic medical records of depression is studied. Time in the text is divided into clear time expression, fuzzy time expression and event-based time expression. Through the study of different time expressions, it is found that both fuzzy time expression and event-based time expression are composed of clear time expression and other words. Since clear time expression is mostly standardized, regular expression is used to recognize it. On this basis, the CRF model and external dictionary features are used to recognize time expression. External dictionary is introduced to improve the accuracy of word segmentation, and time recognition is transformed into sequential tagging. External features are optimized. Event dictionary, time modifier dictionary and time expression dictionary are introduced to solve the problem of time expression location. The results show that the F value of the method of combining regular expression with CRF is 93.33% in the recognition of time expression. In the future, we will continue to study the standardization of time expression and the generation of time line of medical records, which is of great significance for doctors to quickly understand the patient's condition and put forward effective treatment plans.

References

1. Yang, J., Yu, Q., Guan, Y.: Review of named entity recognition and entity relation extraction in electronic medical records. J. Autom. **40**(8), 1537–1562 (2014)
2. Verhagen, M., Sauri, R., Caselli, T.: SemEval-2010 Task 13: TempEval-2. In: Proceedings of the 5th International Workshop on Semantic Evaluation, Stroudsburg, PA, USA, pp. 57–62 (2010)
3. Bethard, S., Derczynski, L., Savova, G.: Semeval-2015 task 6: clinical tempeval. In: Proceedings of the 9th International Workshop on Semantic Evaluation, pp. 806–814 (2015)
4. Wang, Y.: Research on Chinese time information extraction in financial field. Tsinghua University (2004)
5. Zhong, Z., Li, C., Qiao, L.: An efficient method for extracting publishing time of Web news. Minicomput. Syst. **34**(9), 2085–2089 (2013)
6. Zhu, S.S., Liu, Z.T., Fu, J.F., Zhu, F.: Chinese time phrase recognition based on conditional random fields. Comput. Eng. **37**(15), 164–167 (2011)
7. Wu, Q., Huang, D.: Recognition of Chinese time expression based on conditional random field and time lexicon. Chin. J. Inf. Sci. **28**(06), 169–174 + 189 (2014)
8. Liu, L., He, Z., Xing, X.: Chinese time expression recognition based on semantic role. Comput. Appl. Res. **28**(7), 2543–2545 (2011)
9. Jindal, F., Prateek, S., Dan Roth, T.: Extraction of events and temporal expressions from clinical narratives. J. Biomed. Inform. **46**, S13–S19 (2013)
10. Zhou, X.: Time Information Extraction of Chinese Medical Record Text. Zhejiang University (2011)
11. Zhu, L., Yang, H., Yan, Z.: Extracting temporal information from online health communities. In: Proceedings of the 2nd International Conference on Crowd Science and Engineering, pp. 50–55. ACM, Beijing (2017)
12. Ferro, L., Gerber, L., Mani, I.: TIDES 2005 standard for the annotation of temporal expressions. The MITRE Corporation (2005)
13. Lafferty, J., McCallum, A., Pereira, F.C.N.: Conditional random fields: probabilistic models for segmenting and labeling sequence data. In: Proceedings of the 18th International Conference on Machine Learning, Williamstown, pp. 282–289 (2001)
14. Hancks HanLP. https://github.com/hankcs/Hanlp. Accessed 12 Apr 2019

MeKG: Building a Medical Knowledge Graph by Data Mining from MEDLINE

Thuan Pham[1(✉)], Xiaohui Tao[1], Ji Zhang[1], Jianming Yong[2], Xujuan Zhou[2], and Raj Gururajan[2]

[1] School of Sciences, University of Southern Queensland, Toowoomba, Australia
{Thuan.Pham,Xiaohui.Tao,Ji.Zhang}@usq.edu.au
[2] School of Management and Enterprise, University of Southern Queensland, Toowoomba, Australia
{Jianming.Yong,Xujuan.Zhou,Raj.Gururajan}@usq.edu.au

Abstract. Mining data on a knowledge level can help to achieve a higher performance of a decision support system. This study built a knowledge graph based on MEDLINE that has a large number of articles in the medical domain. MEDLINE uses Medical Subject Headings (MeSH) for document index. Based on MeSH, articles are extracted from the MEDLINE correspondent to medical subjects. Using the MeSH as the backbone of knowledge base, the MEDLINE articles were used to generate instances which helped to populate the knowledge base. This approach facilitated the creation of a knowledge graph that was capable of discovering the hidden knowledge among concepts of MeSH. The knowledge graph had a significant effect on improving the quality of healthcare. The contribution of the research is on a framework for building knowledge bases. Moreover, the approach provided an essential source at the knowledge level for researchers in healthcare.

Keywords: MeSH · MEDLINE · Knowledge graph · Data mining

1 Introduction

Increasing amount of big data has introduced many new challenges in data mining. Traditional data mining performed at data level may not be highly effective in discovering knowledge for two reasons. Firstly, each attribute at the data level has a unique label which has a closed world assumption. Secondly, it is difficult to infer implicit information among entities. In contrast, at the knowledge level, each attribute might have more than one label, which focuses on presenting information by semantic meaning rather than data. Therefore, mining at the knowledge level can help infer implicit details which can help to achieve a higher level of knowledge discovery.

In terms of knowledge base, Wang et al. [11] demonstrated that a medical knowledge base would have the ability to improve the performance of discovering medical knowledge if it was integrated into the medical domain for relevance

© Springer Nature Switzerland AG 2019
P. Liang et al. (Eds.): BI 2019, LNAI 11976, pp. 159–168, 2019.
https://doi.org/10.1007/978-3-030-37078-7_16

assessment. Goh et al. [5] argued that a knowledge base is useful in the clinical decision support system. In the medical area, MEDLINE is a vital source because it contains a significant number of articles that are updated every week in the medical field. However, most researchers focus on using MEDLINE to identify information. Xu et al. [13] introduced a model to identify drug-disease associations by extracting the document from MEDLINE. Some researchers have suggested a new method of achieving high quality in discovering knowledge. Banuqitah et al. [1] suggested a way that used more than one level of learning from documents extracted in MEDLINE to improve the discovery of previously hidden precious knowledge. Therefore, MEDLINE would become more useful if it was processed to be integrated into concepts of MeSH by instances which can be applied to applications of decision support.

The study focuses on building a knowledge base to help facilitate the use of MEDLINE as well as to improve the performance of decision support systems in the medical domain based on Medical Subject Headings (MeSH) and MEDLINE. MeSH includes a list of concepts that link to the documents of MEDLINE. Each concept is associated with a number of articles in the medical domain. MEDLINE uses MeSH for indexing of articles. MEDLINE holds approximately 27 million articles in the fields of biomedicine and health [3] and updates frequently. However, researchers cannot directly apply MeSH on data mining because MeSH is a hierarchy of concepts. This study has created feature vectors from documents in MEDLINE and then mapped feature vectors to concepts in MeSH to build a knowledge base, namely **MeKG**. This helps measure the distance among concepts and creates an advantage in using these concepts. The contributions of this work are the following:

- Introducing a framework to build a knowledge base that can help to manage more meaningful information.
- Providing a knowledge base in the medical domain that can improve the result of searching based on semantic relationships among entities.
- Opening a chance for other works to use this knowledge base to help build medical decision support systems.

The remainder of this paper is organised as follows. In Sect. 2, the study reviews the existing work on mining the knowledge base. Then, the study suggests a framework for building a knowledge base in Sect. 3. Section 4 follows, where the study discusses the advantages of the knowledge base. Finally, the conclusions are presented in Sect. 5.

2   Related Work

Knowledge base has become a notable topic in the last decade. It plays a vital role in mining data. It can help to discover hidden patterns between entities. Therefore, there is an increase in building and applying knowledge base on data mining. Xu et al. [12] suggested a new knowledge powered method by incorporating knowledge graphs into the learning process to encode the relationship between entities, attributes or properties of objects. This approach has assisted in improving the quality of word representations. Bordes et al. [2] suggested a method to learn the distributed embedding of knowledge bases. This approach has helped to generate new reasonable relations by linking raw-text as entity vectors to knowledge extraction. Similarity, Nguyen et al. [7] investigated a method to leverage semantics from raw text and knowledge resources for achieving high-level representations of documents based on both text embedding and concept-based embedding.

To use conceptual graphs effectively, Shi et al. [9] proposed a new model to organise and integrate the textual medical knowledge into conceptual graphs. This approach provided semantic mappings between textual medical expertise and medical knowledge, which could explore complex semantics among entities in chain inferences. This proposal helped to detect and obtain access to valuable information from the medical domain. Moreover, based on the documents, Voskarides [10] aimed to clarify the relationships among entities of knowledge graph by sentences. These sentences that referred to an entity pair were extracted and enriched through ranking.

Obviously, knowledge base has been successfully considered for using in data mining by researchers [2,7,9,10,12]. However, researchers had not previously considered developing a framework of building a knowledge graph that helps to improve the quality of decision support system. Besides, a useful source, MEDLINE that contains a large number of articles in the medical domain was not fully explored. This source would be significant for healthcare if it could be processed for applying in data mining. Therefore, this study presents a framework to build a knowledge graph based on MEDLINE to improve the effect of exploring knowledge in the medical domain. In addition, this approach aims to help increase the accuracy of decision support systems.

3   Building Knowledge Base

3.1   The Framework

The study aims to connect specific factors to concepts for generating instances, which help to identify the distance between concepts. For example, it is impossible to calculate the semantic distance between Google and Java because Google and Java are not concepts. However, by applying a training model, researchers are able to recognize that Google is an instance of a search engine, and Java is an instance of programming. As a result, the researchers can measure the distance between Java and Google by calculating the passages existing between concepts within the graph. Therefore, this study proposes an approach to improve the quality of healthcare by linking feature vectors (just like instances) to concepts which can help to find more relationships between concepts. To deal with this challenge, the research grouped all the kinds of concepts into a subgraph through medical subject headings. Each subgraph corresponds to a specific disease or a subject. Then, the study populated the knowledge base for each subgraph based on instances that learned by a large number of articles from MEDLINE. These instances were mapped to concepts in MeSH to create the **MeKG** knowledge graph base. The **MeKG** can make a significant contribution to decision-making support which helps to find other options in term of possible medication and diagnoses for practitioners.

However, MeSH is one of the hierarchy of those concepts and MEDLINE is an aggregation document in the medical domain as it collects published papers. MEDLINE is a metadata collection repository of biomedical abstracts and one of the most significant data sources related to scientific literature. MEDLINE uses MeSH to manually index publications from the National Library of Medicine. Currently, MEDLINE holds approximately 27 million records in the fields of biomedicine and health [3]. This study did not consider citations, including the links of articles as well as book reviews. The study used only journal articles that were indexed by MeSH to perform experiments. There are six to fifteen subject headings from MeSH assigned with each article in the MEDLINE database. Using MeSH, which stores concepts from documents in MEDLINE, has an advantage in building a knowledge base. In this case, MeSH plays an important role as the backbone of the graph for building the **MeKG** knowledge graph base. This study applied instances that were learned from MEDLINE to populate knowledge base constructed on the following formal definition.

Definition 1. *[Medical Subject Headings]*
The Medical Subject Headings are $\mathbb{C} = \{c_1, c_2, \ldots, c_i\}$, *where* c *is a concept belong* $\mathbb{C}$ *and* i *is the number of concepts.*

Definition 2. *[Knowledge Graph Base]*
The Knowledge Graph Base is a 3-tuple $\mathbb{KG} := \langle \mathcal{G}, \mathcal{R}, \mathbf{G}_{\mathcal{G}}^{\mathcal{R}} \rangle$, *where*

- $\mathcal{G} := \langle \mathbb{C}, \mathcal{I} \rangle$, $\epsilon \in \mathcal{G}, \epsilon := \langle c, \mathcal{I}_c \rangle$, *where* $\mathcal{I}_c \subset \mathcal{I}, c \in \mathbb{C}$. $\mathcal{I}$ *is the universal set of instances.*
- $\mathcal{R} = \{r_1, r_2, \ldots, r_q\}$ *is the set of all relation types in a knowledge graph, where* q *is the number of relations.*
- $\mathbf{G}_{\mathcal{G}}^{\mathcal{R}}$ *is graph that is generated by* $\mathcal{R}$ *and* $\mathcal{G}$.

Definition 3. *[MEDLINE]*
The MEDLINE is $\mathbb{D} = \{d_1, d_2, \ldots, d_j\}$, *where* j *is the number of documents in MEDLINE.* $d := \langle \mathcal{T}, map(d) \rangle$, *where* $\mathcal{T} = \{t_1, t_2, \ldots, t_z\}$ *is a set of terms from* d, *and* $map(d) \longrightarrow \mathbb{C}_d \subset \mathbb{C}$.

Definition 4. *[Research Problem]*
The task of research is to learn instances from $\mathbb{D}$ *based on* $\mathcal{T}$ *and* $map(d)$. *These instances then are associated with concepts* $\mathbb{C}_d \subset \mathbb{C}$ *to build a knowledge graph* $\mathbb{KG}$.

The challenge of this study was related to learning instances from $\mathbb{D}$ which store a large number of articles in different subjects in the medical domain. Therefore, this study divided the documents from $\mathbb{D}$ into different subjects based on topics from $\mathbb{C}$ with the use of concepts. Each concept c in MeSH will belong to a specific subject. Assume that there were n subjects which contain k number of concepts. In this case, MeSH was presented as $\mathbb{C} = \{c_{11}, c_{22}, \ldots, c_{kn}\}$. MEDLINE was indicated as $\mathbb{D} = \{d_{j1}, d_{j2}, \ldots, d_{jn}\}$. This study took the advantage of $\mathbb{C}$ and $\mathbb{D}$ that they were linked through descriptors of MeSH. Each article in $\mathbb{D}$ has three to six concepts of $\mathbb{C}$ which belong to a specific subject. Based on those concepts, a large number of articles would be extracted. The extracted documents were used to learn instances from those concepts.

Assume that we want to learn instances from subject $s \in n$, we have $\mathcal{C}_s :=$ $\langle \mathbb{C}_s, \mathcal{I}_s \rangle$, where $\mathcal{C}_s \subset \mathcal{C}$. $\mathbb{C}_s = \{c_{1s}, c_{2s}, \ldots, c_{ks}\}$. $\mathbb{D}_s = \{d_{js}, d_{js}, \ldots, d_{js}\}$
$\mathcal{I}_s$ is learned by mapping between $\mathbb{C}_s$ and $\mathbb{D}_s$.

$$f(\mathcal{I}_s) = \sum_{k=1}^{l} (c_{ks} \longmapsto t_{zs}) \times \alpha \qquad (1)$$

where α is a threshold to determine the mapping

3.2   Building a Knowledge Graph Base

Before using MeSH and MEDLINE for building a knowledge graph, the study needs to rebuild the XML format of MeSH and MEDLINE. Figure 1 presents some important elements of structure MEDLINE and MeSH. In contrast, Figs. 2 and 3 showed a new format of MeSH by tables that were stored in the form of MySQL. The new format can help to provide efficient access to the concepts and relationships stored in MeSH. The study used Java programming language to write an XML parser for converting the data. The new structure can be a more convenient way to extract information from MEDLINE, which uses MeSH for information retrieval.

By rebuilding the structure of MeSH and MEDLINE, MeSH and MEDLINE can be easily accessed to extract all the articles related to specific fields. The field depends on the purpose of extraction data. Based on MeSH, there are many types of topics in this graph. Each subject may have several different objects. The object with the same type may have a list of a number of concepts and terms. However, MeSH is still a hierarchy of concepts and terms, and it is not valid to use directly for discovery knowledge and mining data. Therefore, this approach makes a task to create instances that link to these concepts from MeSH, which help to find semantic distance between concepts. The study uses a word vector space to create features vectors that correspond to instances. Word vector space called word embedding was approached for calculating the weight of each concept and term. Word2Vec is a successful algorithm related to word vector space for generating feature vectors indicated by Mikolov et al. [6]. This technique is used by Ganguly et al. [4] and Zheng et al. [14] in measuring the semantic similarity among documents. In this study, Word2Vec is also used for learning instances.

Assume we want to find the relationships among subjects about the heart disease. We select all the articles related to the heart from MEDLINE based

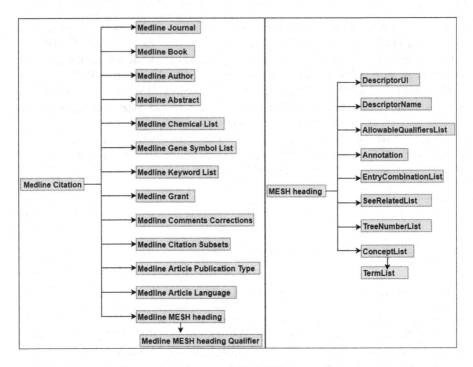

Fig. 1. Attributes of MEDLINE and MeSH

iddesqua	DescriptorUI	QualifierUI
1	D000001	O000031.O000276.O000493.O000096.O00062...
2	D000002	O000031.O000276.O000493.O000627.O00063...
3	D000003	O000145.O000191.O000266.O000331.O00036...
4	D000005	O000000981.O000002.O000033.O000098.O00...
5	D000006	O000469.O000276.O000382.O000175.O00000...
6	D000007	O000469.O000276.O000382.O000175.O00000...
7	D000008	O000469.O000276.O000382.O000175.O00000...
8	D000009	O000469.O000002.O000276.O000382.O00000...
9	D000010	O000469.O000002.O000276.O000382.O00000...
10	D000011	O000145.O000187.O000201.O000235.O00025...
NULL	NULL	NULL

Fig. 2. A table extracted from MeSH for Descriptor link to Qualifier

on descriptors of the MeSH. In this case, the descriptor represents heart disease in MeSH through an identity (D006321). Figure 4 shows the associations between MeSH and MEDLINE. Based on the identity of the description, the study performs an extraction of all articles related to heart disease. These documents have a list of terms regarding heart disease. To ensure enough data for calculating the weight of words, the research cleaned data by removing all stop-words and steam-words. Then, the study used the word2vec algorithm [6]

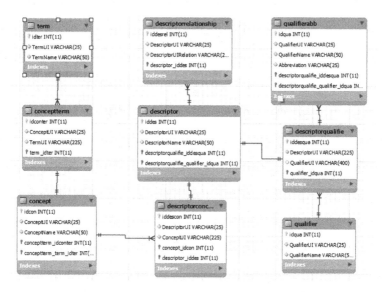

Fig. 3. The relationship among tables extracted from MeSH

to convert the extracted data into vector space. In this experiment, the study used both two methods, including the continuous bag of words and the skip-gram to train the model. The skip-gram was selected for the final generation of feature vectors because the model using the skip-gram obtained a better result than the continuous bag of words. The set of parameters for the skip-gram method to achieve the highest performance for this experiment including set_Min_Vocabulary_Frequency (5), use_Number_Threads (20), set_Window_Size (10), set_Layer_Size (200), use_Negative_Samples (10) and set_Number_Iterations (5). This process helped to calculate the semantic relationship between terms related to heart disease and to identify similar neighbours for a given word. The weight among terms was determined by a coefficient α rank from 1 to 0. Finally, all coefficient α learned from this training which are called instances would play an essential role in populating the knowledge for the graph. Instances can link to concepts from MeSH through mapping between these selected features and objects for heart disease. The mapping between features and concepts help to create a knowledge graph which assists in finding hidden meaning between concepts. This knowledge graph was presented by Fig. 5.

4   Discussions

There is an increasing amount of research in healthcare by mining data at the data level. Based on these methods, all hidden relationships among objects may not be fully explored because of the ambiguous meaning of objects. To improve the performance of a decision support system and promote it to the knowledge level, the study suggested a method to create a knowledge base with a focus on

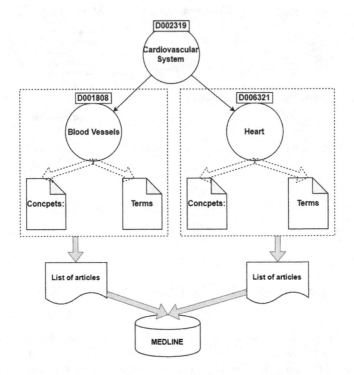

Fig. 4. Mapping between MeSH and MEDLINE

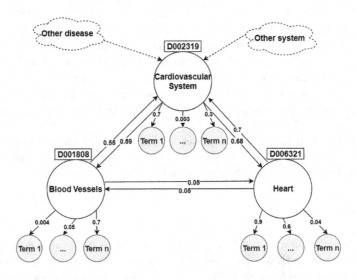

Fig. 5. Knowledge Graph with instances by learning from MEDLINE

searching for the precise meaning of each concept as well as hidden relationships between concepts. For example, if the data level is used to find relationships between smoking and lung cancer, the researcher may not be able to work out how smoking causes cancer. However, using instances that are learned from the knowledge graph may help researchers understand how lung cancer causes smoking. Based on the knowledge graph, the similarity and difference among objects can be presented under the weights. This advantage assists in achieving a higher performance of mining medical and healthcare data.

Specifically, knowledge base has strong capabilities to improve the performance of the classification models. For example, if the study is only based on the data level to build the classification model, this approach may not create a useful model because of the semantic relationships among objects. However, by using the knowledge level, these issues would be resolved. Knowledge base may help to discover more unknown objects that can help the classification model to be able to generate an effective result. At the data level, the classification model has a trend for using all attributes of a dataset to predict results [8]. However, at the knowledge level, the classification model can reject all attributes that are not related to the topic for predicting a result based on the semantic relationships among attributes of a dataset. In addition, eliminating noise variables or noise properties helps to improve the accuracy of the result and plays a significant role in classification models. Therefore, application of a knowledge base in developing classification models promises to bring positive results.

5   Conclusions

Mining data at the data level has challenges because of ambiguous meaning and an increase volume in big data. In contrast, mining data at the knowledge level may help to discover relationships, which are hidden between entities. Therefore, this study introduced a method to build the **MeKG** knowledge graph base that helped improve the performance of decision support systems. A vector space model was conducted to generate feature vectors for linking to concepts of MeSH. The model runs all documents extracted from MEDLINE for a specific topic. These feature vectors helped to create instances. Finally, instances were mapped to concepts and terms of MeSH for building the **MeKG** knowledge graph base. This study contributed a significant knowledge graph base in healthcare. Additionally, this study helped medical researchers, as well as practitioners, achieve a high performance of searching based on a knowledge base level. The **MeKG** knowledge graph base also played an essential role in the searching-system because it could help to solve the ambiguous meaning of each object.

In further research, the study aims to use this knowledge base for applying the classification model to improve its predicted capability.

References

1. Banuqitah, H., Eassa, F., Jambi, K., Abulkhair, M.: Two level self-supervised relation extraction from MEDLINE using UMLS. Int. J. Data Min. Knowl. Manag. Process **6**(3), 11–23 (2016)
2. Bordes, A., Weston, J., Collobert, R., Bengio, Y.: Learning structured embeddings of knowledge bases. In: Twenty-Fifth AAAI Conference on Artificial Intelligence (2011)
3. Costa, J.P., et al.: Mining MEDLINE for the visualisation of a global perspective on biomedical knowledge. In: KDD 2018 (24th ACM SIGKDD Conference on Knowledge Discovery and Data Mining) (2018)
4. Ganguly, D., Roy, D., Mitra, M., Jones, G.J.: Word embedding based generalized language model for information retrieval. In: Proceedings of the 38th International ACM SIGIR Conference on Research and Development in Information Retrieval, pp. 795–798. ACM (2015)
5. Goh, W.P., Tao, X., Zhang, J., Yong, J.: Decision support systems for adoption in dental clinics: a survey. Knowl.-Based Syst. **104**, 195–206 (2016)
6. Mikolov, T., Sutskever, I., Chen, K., Corrado, G.S., Dean, J.: Distributed representations of words and phrases and their compositionality. In: Advances in Neural Information Processing Systems, pp. 3111–3119 (2013)
7. Nguyen, G.-H., Tamine, L., Soulier, L., Souf, N.: Learning concept-driven document embeddings for medical information search. In: ten Teije, A., Popow, C., Holmes, J.H., Sacchi, L. (eds.) AIME 2017. LNCS (LNAI), vol. 10259, pp. 160–170. Springer, Cham (2017). https://doi.org/10.1007/978-3-319-59758-4_17
8. Pham, T., Tao, X., Zhanag, J., Yong, J., Zhang, W., Cai, Y.: Mining heterogeneous information graph for health status classification. In: 2018 5th International Conference on Behavioral, Economic, and Socio-Cultural Computing (BESC), pp. 73–78. IEEE (2018)
9. Shi, L., Li, S., Yang, X., Qi, J., Pan, G., Zhou, B.: Semantic health knowledge graph: semantic integration of heterogeneous medical knowledge and services. BioMed Res. Int. **2017**, 12 (2017)
10. Voskarides, N., Meij, E., Tsagkias, M., De Rijke, M., Weerkamp, W.: Learning to explain entity relationships in knowledge graphs. In: Proceedings of the 53rd Annual Meeting of the Association for Computational Linguistics and the 7th International Joint Conference on Natural Language Processing (Volume 1: Long Papers), pp. 564–574 (2015)
11. Wang, H., Zhang, Q., Yuan, J.: Semantically enhanced medical information retrieval system: a tensor factorization based approach. IEEE Access **5**, 7584–7593 (2017)
12. Xu, C., et al.: RC-NET: a general framework for incorporating knowledge into word representations. In: Proceedings of the 23rd ACM International Conference on Information and Knowledge Management, pp. 1219–1228. ACM (2014)
13. Xu, R., Wang, Q.: Large-scale extraction of accurate drug-disease treatment pairs from biomedical literature for drug repurposing. BMC Bioinform. **14**(1), 181 (2013)
14. Zheng, G., Callan, J.: Learning to reweight terms with distributed representations. In: Proceedings of the 38th International ACM SIGIR Conference on Research and Development in Information Retrieval, pp. 575–584. ACM (2015)

Specificity Analysis of Picture-Induced Emotional EEG for Discrimination Between Schizophrenic and Control Participants

Hongzhi Kuai[1,2,5], Yang Yang[2,4,5], Jianhui Chen[2,3,5], Xiaofei Zhang[2,3], Jianzhuo Yan[3], and Ning Zhong[1,2,3,5(✉)]

[1] Department of Life Science and Informatics, Maebashi Institute of Technology, Maebashi, Gunma 371-0816, Japan
{m1956503,zhong}@maebashi-it.ac.jp
[2] International WIC Institute, Beijing University of Technology, Beijing, China
[3] Faculty of Information Technology, Beijing University of Technology, Beijing, China
[4] Department of Psychology, Beijing Forestry University, Beijing, China
[5] Beijing International Collaboration Base on Brain Informatics and Wisdom Services, Beijing, China

Abstract. Emotion processing, playing an important role in our social interactions, is a sub-topic of social cognition. Significant differences in emotion perception and processing have been demonstrated between schizophrenia and normal people. Therefore, it is a very effective strategy to use the emotional stimulation as the core means to explore the difference between patients and normal people, and then to develop the discriminative model for patients with schizophrenia. In this paper, emotional images were used to stimulate the two groups (schizophrenia group and control group), and the electrophysiological signals during the experiment were recorded. In the feature extraction phase, the time-domain dynamics and the asymmetry of the hemisphere were considered at different stimulation stages. Finally, five effective machine learning methods were used to distinguish between schizophrenia and healthy controls under positive and negative emotional stimuli, respectively. The experimental results show that the two groups of event-related electrophysiological signals obtained by negative stimulation can be better distinguished than those obtained by positive stimulation. And, this phenomenon is more pronounced in the time window of first second after the stimulus appears. Meanwhile, the highest average F-score with 10-fold cross-validation strategy can reach 0.994 by combining both support vector machine classifier and grid search methods.

Keywords: Schizophrenia · Emotion · Machine learning · Prediction · Electroencephalography

© Springer Nature Switzerland AG 2019
P. Liang et al. (Eds.): BI 2019, LNAI 11976, pp. 169–178, 2019.
https://doi.org/10.1007/978-3-030-37078-7_17

1   Introduction

Social cognition refers to mental operations involved in the processing of social cues and includes the domains of emotion processing, theory of mind, social perception, social knowledge and attributional bias [1]. Impaired social cognition and skills in schizophrenia not only bring heavy mental pressure to themselves and their family, but also bring a lot of economic burden and security threats to society. Therefore, effective recognition for schizophrenia patients are particularly important.

The diagnosis of schizophrenia primarily depends on the clinician's evaluation based on previous medical records, corroborative information from caregivers, mental state examination and so on. In this process, the patient's clinical symptoms and behavioral signs as an important basis are also identified and assessed through the interview with a psychiatrist. Meanwhile, some diagnostic taxonomies guidelines as the standard, such as the Diagnostic and Statistical Manual of Mental disorders, Fifth Edition (DSM-5) [2] and International Classification of Diseases, 10th revision (ICD-10) [3], are used to judge the type of disease and the degree of illness of the patient. Obviously, this process is deeply dependent on the individual experience of the clinician. Throughout the years, extensive works have been dedicated to optimizing the process of diagnosis and treatment in schizophrenia, such as genetic studies [4], objective biomarkers from MRI [5], fMRI [6], and Electroencephalography (EEG) [7,8], etc. Especially, objective quantitative indicators for diagnosing schizophrenia are highly desirable to supplement clinical efforts. Furthermore, there has been increasing interest for the reliability of recognition in current psychiatric practice. Some theory-driven methods with psychophysiological behavioral inspiration [9] and data-driven methods with machine learning [10] are increasingly being used for the auxiliary diagnosis and identification of schizophrenia.

This paper mainly focuses on the differences between schizophrenia and control groups at different stimulation stages during emotional picture viewing. The EEG signal is used as a tool to decode brain state and its effectiveness has been widely proven. In the pattern recognition phase, multiple classifiers with EEG features of statistical properties and hemispheric asymmetry are designed to maximize the distinction between patients with schizophrenia and normal people for positive and negative stimuli, respectively. The remainder of this paper is structured as follows: a detailed description of the experimental setup and the methodology is presented in Sect. 2; the classification results and evaluation are presented in Sect. 3; and the conclusion is given in Sect. 4.

2   Experiment and Methods

2.1   Stimuli

Affective pictures were selected from the Chinese Affective Picture System (CAPS), which consisted of 852 selected images, and 46 Chinese university students were collected to the valence, arousal and dominance by using of

Self-Assessment Manikin with a 9-point scale [11]. Similar to the International Affective Picture System [12], the scope of CAPS is as wide-ranging as possible and introduces some scenes involving oriental colors. In this experiment, each participant watched 120 pictures in which three categories of pictures (40 for positive, 40 for negative, and 40 for natural) were selected based on normative ratings in valence and arousal. The contents of positive pictures included beautiful scenery, happy children, romance, and gift, with mean valence of 6.88 ± 0.28 and mean arousal of 5.97 ± 0.42. Negative pictures, such as conflict, attacks, and dead bodies were adopted for the negative stimuli, with mean valence of 2.73 ± 0.34 and mean arousal of 4.92 ± 0.29. Pictures of sports scene and natural environment were used as the neutral stimuli, with mean valence of 4.89 ± 0.83 and mean arousal of 5.16 ± 0.46.

2.2   Procedure

The event-related or single-trial EEG experiments induced by emotional picture are designed. And all pictures were randomly presented to the participants on every run. Firstly, the participants performed some practice trials to familiarize themselves with the system. Some pictures were shown during the unrecorded trials. Next, the operators left the room and started the EEG signal recording. Then, the 120 pictures arounded by red or blue boxes were presented. Each trial consists of the following steps (see in Fig. 1): (1) 500 ms display of the fixation cross to inform the participants of their progress, (2) 500 ms of rest time, (3) Display of the pictures arounded by red or blue boxes, (4) 0–3000 ms of response time for judging the color of the box by pressing two key on the keyboard, (5) 2000 ms of self-elicitation of emotion, during which participants watch the scene of black background with an asterisk in the middle, and (6) 500 ms of rest time.

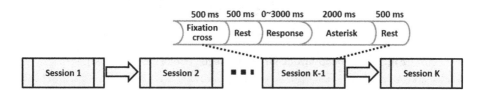

Fig. 1. Protocol of the event-related EEG experimental design.

2.3   Data Acquisition and Preprocessing

For this study, the recruitment of schizophrenia participants was approved by the Ethics Committee of Suzhou Guangji Hospital, Suzhou. Each participant was required to sign a written informed consent document before participating in the study. We informed them of the purpose and procedure of the experiment, and that the physiological recording device was harmless.

Data Acquisition. We designed an experiment and recorded the EEG signals of participants (10 males, 5 for patients with schizophrenia and 5 for normal people). All participants (age from 18 to 55) were selected through interviews and clinically assessments based on DSM-5. And all participants were right-handed with normal vision and hearing.

All stimuli were displayed on the center of the screen to minimize eye movement. The software E-Prime 3.0 (Psychology Software Tools, Sharpsburg, GA, USA) was used to present stimuli, mark synchronization labels, and record the participants' responses. BrainVision Recorder Professional (Brain Products GmbH, Gilching, Germany) software version 1.21.0102 was used for data recording. The layout of 32 electrodes followed the international 10–20 system, as shown in Fig. 2.

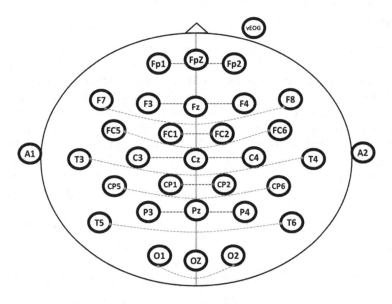

Fig. 2. EEG cap layout for 29 channels. (Color figure online)

One vertical EOG (vEOG) channel was treated like EEG channels, hence, EEG settings also apply to EOG recordings. The A1 and A2 electrodes were used for reference calculation. Thus, the number of effective EEG electrodes was 29 in Fig. 2.

Data Preprocessing. To ensure that all EEG samples had the same length, we took the last 1000 ms before the picture stimulus appeared and the 3000 ms after the picture was presented for analysis. The raw signals were reduced the sampling rate to 1000 Hz. Then, high-frequency interferences were filtered out from EEG signals by using a band-pass filter with a range of 0.1–45 Hz. Finally, all EEG signals were reset by using of reference electrodes, A1 and A2.

2.4   Methods

In this study, each emotional EEG sample with a duration of 4000 ms was segmented by using a 1000 ms window and no overlap between two consecutive
windows. Then, the EEG signals of different time windows were separately analyzed, such as 1000 ms before stimulation as stage 1 and 0–1000 ms as stage 2,
1000–2000 ms as stage 3, 2000–3000 ms as stage 4 after stimulation. Therefore,
we acquired 120 labeled samples (40 for positive, 40 for negative and 40 for neutral) for each participant who watched 120 pictures in each stage. And, each type
of emotional stimulus corresponds to 400 samples (200 for schizophrenia and 200
for normal people) with respect to each stage. Currently, we only focus on the
physiological responses induced by positive and negative stimuli. Two strategies
of feature extraction were utilized for discrimination between schizophrenics and
control participants during emotional picture viewing: intra- and inter-channel
EEG patterns based discrimination. Finally, the obtained features were fed into
multiple classifiers to achieve effective discrimination between schizophrenics and
control participants.

Feature Extraction. As mentioned above, two feature extraction strategies are
considered in this study. On the one hand, Hjorth Parameters (Activity, Mobility
and Complexity), introduced by Hjorth in 1970 [13], were used to characterize
the EEG patterns in terms of amplitude, time scale and complexity [14]. Three
parameters are indicators of statistical properties used in the analysis of EEG
signals for signal processing in the time domain, which are defined as follows.

The *Activity* measures the degree of deviation of the signal amplitude:

$$Activity = \frac{1}{N} \sum_{n=1}^{N} (s(n) - \mu_s)^2 \tag{1}$$

The *Mobility* measures the change in slope:

$$Mobility = \sqrt{\frac{var(s'(n))}{var(s(n))}} \tag{2}$$

The *Complexity* measures how many standard slopes are on one amplitude:

$$Complexity = \frac{mobility(s'(n))}{mobility(s(n))} \tag{3}$$

where μ_s is mean value of the signal s, N is the sample number of the signal s,
$s'(n)$ is first derivative of the signal s, and $var(s)$ represents the variance of the
signal s.

On the other hand, hemispheric asymmetry as an important mechanism has
been widely used in a variety of disease research, including schizophrenia. For
example, Barnett et al. studied interhemispheric transfer times between the
hemispheres, which were assessed by using a lateralized lexical-decision task

in males with schizophrenia and matched controls [15]. In Fig. 2, twelve pairs of electrodes were obtained based on the yellow vertical line in the middle as the axis of symmetry. The two electrodes connected by the green line are electrode pairs, and the two electrodes marked by the red line are the same. So, we can get the left side electrodes of all electrode pairs in L and right side electrodes in R, which are expressed as L = {FP1, F3, C3, FC1, FC5, PC1, PC5, P3, O1, F7, T3, T5} and R = {FP2, F4, C4, FC2, FC6, PC2, PC6, P4, O2, F8, T4, T6}. Then, we calculated three asymmetry indicators, including differential asymmetry (DASM), rational asymmetry (RASM) and asymmetry coefficient (CASM) [16,17]. A detailed description is as follows:

$$DASM = Feature(s_L) - Feature(s_R) \tag{4}$$

$$RASM = \frac{Feature(s_L)}{Feature(s_R)} \tag{5}$$

$$CASM = \frac{Feature(s_L) - Feature(s_R)}{Feature(s_L) + Feature(s_R)} \tag{6}$$

where s_L and s_R represent the EEG signals collected by the left and right electrodes, respectively. And $Feature(s)$ represents the features extracted on the signal s. In this study, the calculation of asymmetry indicators are based on the previously single channel features calculated by Hjorth Parameters.

Classification Methods. Machine learning has been applied to many clinically relevant problems, such as automatic diagnosis, prediction of treatment outcomes and longitudinal illness course, drug evaluation and treatment selection, etc. In this study, several commonly machine learning methods were selected to distinguish schizophrenia and control groups, including k-nearest neighbors algorithm (k-NN), support vector machines (SVM), multilayer perceptron (MLP), Decision Trees (DTs) and Random forests (RFs). The grid search with 10-fold cross validation is used to select the best hyperparameters to employ in the final selected model.

Performance Evaluation. The $F1 - score$ is used to measure test's accuracy, which seek a balance between $Precision$ and $Recall$ to evaluate the validity of the results more reasonably. The formula for $F1 - score$ is

$$F1 - score = 2 * \frac{Precision * Recall}{Precision + Recall} \tag{7}$$

where $Precision = \frac{tp}{tp+fp}$, $Recall = \frac{tp}{tp+fn}$, tp is the number of samples whose label is positive and is predicted to be positive, fp is the number of samples whose label is negative and is predicted to be positive, fn is the number of samples whose label is positive and is predicted to be negative.

3   Results and Evaluation

3.1   Intra-channel EEG Pattern Based Results and Evaluation

In this part of the scenario, the features of the EEG signals are extracted by calculating the Hjorth Parameters for each channel. Then, the classification of the samples from schizophrenia and normal people for four stages of positive picture stimulation was performed. The average F1-score with its standard deviation was obtained from the recognition results of different classifiers, as shown in Table 1. And, the classification performance of four stages with negative picture stimulation is demonstrated in Table 2.

Table 1. Average F1-score with its standard deviations for the discrimination of schizophrenics from control participant under positive stimulation.

Classifiers	Stage 1	Stage 2	Stage 3	Stage 4
KNN	0.950 ± 0.0471	0.945 ± 0.0354	0.950 ± 0.0298	0.940 ± 0.0548
SVM	0.962 ± 0.0365	0.963 ± 0.0306	0.979 ± 0.0233	0.976 ± 0.0222
MLP	0.975 ± 0.0190	0.965 ± 0.0337	0.971 ± 0.0370	0.978 ± 0.0274
DTs	0.896 ± 0.0460	0.913 ± 0.0579	0.894 ± 0.0519	0.890 ± 0.0298
RFs	0.953 ± 0.0422	0.950 ± 0.0298	0.955 ± 0.0384	0.951 ± 0.0311

Table 2. Average F1-score with its standard deviations for the discrimination of schizophrenics from control participant under negative stimulation.

Classifiers	Stage 1	Stage 2	Stage 3	Stage 4
KNN	0.965 ± 0.0284	0.963 ± 0.0287	0.936 ± 0.0324	0.950 ± 0.0350
SVM	0.972 ± 0.0169	0.994 ± 0.0126	0.985 ± 0.0158	0.975 ± 0.0190
MLP	0.975 ± 0.0190	0.986 ± 0.0190	0.977 ± 0.0221	0.976 ± 0.0341
DTs	0.891 ± 0.0348	0.899 ± 0.0672	0.944 ± 0.0263	0.872 ± 0.0397
RFs	0.956 ± 0.0347	0.959 ± 0.0247	0.943 ± 0.0287	0.959 ± 0.0307

Combining the results from Tables 1 and 2, we found that the recognition performance of SVM and MLP are significantly better than other classifiers. And the recognition results corresponding to negative stimulus were better than that of positive stimulus. In particular, the highest F1-score of 0.994 was got in the early stage of negative emotion processing induced by pictures.

Further, we compare the recognition performance of four stages for these two classifiers under different emotional stimulation conditions. For the condition of positive emotional stimulation, we can find that the recognition results of four stages for each classifier are similar from Table 1. For the condition of negative emotional stimulation, however, we can find that the recognition results in Stage 2 and Stage 3 are better than that other stages from Table 2.

To comprehensively concluded our results, it could be regarded that the responses of schizophrenia patients to emotional stimuli are different from that of normal people, especially in the early stages of negative emotion processing. The results are in line with previous studies [18,19].

3.2   Inter-channel EEG Pattern Based Results and Evaluation

In this part of the scenario, the pattern of the EEG signals is extracted by calculating the asymmetry indicators for each pair of symmetrical channels. The classification performances for positive and negative picture-stimulated EEG experiments are shown in Tables 3 and 4, respectively.

Table 3. Average F1-score with its standard deviations for the discrimination of schizophrenics from control participant under positive stimulation.

Classifiers	Stage 1	Stage 2	Stage 3	Stage 4
KNN	0.836 ± 0.0532	0.794 ± 0.0546	0.861 ± 0.0703	0.826 ± 0.0479
SVM	0.929 ± 0.0513	0.934 ± 0.0435	0.937 ± 0.0508	0.918 ± 0.0385
MLP	0.938 ± 0.0496	0.936 ± 0.0467	0.940 ± 0.0365	0.927 ± 0.0267
DTs	0.773 ± 0.0595	0.836 ± 0.0589	0.796 ± 0.0350	0.775 ± 0.0384
RFs	0.898 ± 0.0632	0.897 ± 0.0583	0.882 ± 0.0649	0.888 ± 0.0365

Table 4. Average F1-score with its standard deviations for the discrimination of schizophrenics from control participant under negative stimulation.

Classifiers	Stage 1	Stage 2	Stage 3	Stage 4
KNN	0.789 ± 0.0789	0.836 ± 0.0272	0.851 ± 0.0592	0.839 ± 0.0728
SVM	0.924 ± 0.0587	0.913 ± 0.0452	0.918 ± 0.0368	0.920 ± 0.0298
MLP	0.897 ± 0.0512	0.921 ± 0.0425	0.933 ± 0.0337	0.947 ± 0.0432
DTs	0.760 ± 0.0668	0.805 ± 0.0629	0.806 ± 0.0638	0.764 ± 0.0479
RFs	0.872 ± 0.0618	0.910 ± 0.0558	0.887 ± 0.0529	0.889 ± 0.0351

From Tables 3 and 4, we can find that the SVM and MLP classifiers still show better recognition performances among these five classifiers, although the current results are not as good as those in the previous section. And, the differences in recognition performance among the various classifiers become greater. Moreover, the recognition results corresponding to the four stages are very similar for SVM and MLP classifiers. For the other three classifiers, however, we can find the significant differences of recognition results among the four stages during negative emotional experience. These conclusions about the difference of emotional responses in the early stages of negative stimulation are still established here.

4   Conclusion

Abnormal perception and response to emotion are widespread in patients with schizophrenia, which can significantly impact socio-occupational ability. Studying the differences between schizophrenia and normal people from machine learning with emotional perspective, which is beneficial to improve the diagnosis and evaluation of schizophrenia. This paper explores the differences between schizophrenia patients and normal people under emotional picture stimulation by using of multiple machine learning algorithms. From the experimental results, we can see that the differences from the picture induced emotion signature between schizophrenia patients and normal people are existed, and this differences become more pronounced in the early post-stimulation (e.g., 0–2000 ms after stimulation). Moreover, negative pictures are easier to stimulate differences between the two groups. In future work, the early-stage physiopsychological mechanism of schizophrenia patients to negative picture stimuli will be further explored and excavated.

Acknowledgements. This work was supported by grants from the National Natural Science Foundation of China (61420106005), the Science and Technology Project of Beijing Municipal Commission of Education (KM201710005026), the JSPS Grants-in-Aid for Scientific Research of Japan (19K12123), and the Web Intelligence Consortium (WIC).

References

1. Thonse, U., Behere, R.-V., Frommann, N., et al.: Social cognition intervention in schizophrenia: description of the training of affect recognition program-Indian version. Asian J. Psychiatry **31**, 36–40 (2018)
2. American Psychiatric Association: Diagnostic and Statistical Manual of Mental Disorders. Fifth Edition (DSM-V). American Psychiatric Publishing, Philadelphia (2013)
3. World Health Organization: International Classification of Diseases (ICD-10). World Health Organization, Geneva (1992)
4. Franke, B., et al.: Genetic influences on schizophrenia and subcortical brain volumes: large-scale proof of concept. Nature Neurosci. **19**(3), 420 (2016)
5. Xiao, Y., Yan, Z., Zhao, Y., et al.: Support vector machine-based classification of first episode drug-naïve schizophrenia patients and healthy controls using structural MRI. Schizophr. Res. (2017)
6. Wang, S., et al.: Abnormal regional homogeneity as a potential imaging biomarker for adolescent-onset schizophrenia: a resting-state fMRI study and support vector machine analysis. Schizophr. Res. **192**, 179–184 (2018)

7. Johannesen, J.K., Bi, J., Jiang, R., et al.: Machine learning identification of EEG features predicting working memory performance in schizophrenia and healthy adults. Neuropsychiatric Electrophysiol. **2**(1), 3 (2016)

8. Grin-Yatsenko, V.-A., Ponomarev, V.-A., Pronina, M.-V., Poliakov, Y.-I., Plotnikova, I.-V., et al.: Local and widely distributed EEG activity in schizophrenia with prevalence of negative symptoms. Clin. EEG Neurosci. **48**(5), 307–315 (2017)

9. Maia, T.V., Huys, Q.J.M., Frank, M.J.: Theory-based computational psychiatry. Biol. Psychiatry **82**(6), 382–384 (2017)

10. Guloksuz, S., Rutten, B.P.F., Pries, L.K., et al.: The complexities of evaluating the exposome in psychiatry: a data-driven illustration of challenges and some propositions for amendments. Schizophr. Bull. **44**(6), 1175–1179 (2018)

11. Bai, L., Ma, H., Huang, Y.-X., et al.: The development of native Chinese affective picture system-a pretest in 46 college students. Chin. Ment. Health J. **19**(11), 719–722 (2015)

12. Lang, P.-J., Bradley, M.-M., et al.: International affective picture system (IAPS): instruction manual and affective ratings. Center for Research in Psychophysiology, University of Florida, Gainesville (2001)

13. Hjorth, B.: EEG analysis based on time domain properties. Electroencephalogr. Clin.Neurophysiol. **29**(3), 306–310 (1970)

14. Oh, S.-H., Lee, Y.-R., et al.: A novel EEG feature extraction method using Hjorth parameter. Int. J. Electron. Electr. Eng. **2**(2), 106–110 (2014)

15. Barnett, K.-J., Kirk, I.-J.: Lack of asymmetrical transfer for linguistic stimuli in schizophrenia: an ERP study. Clin. Neurophysiol. **116**(5), 1019–1027 (2005)

16. Duan, R.-N., Wang, X.-W., Lu, B.-L.: EEG-based emotion recognition in listening music by using support vector machine and linear dynamic system. In: Huang, T., Zeng, Z., Li, C., Leung, C.S. (eds.) ICONIP 2012. LNCS, vol. 7666, pp. 468–475. Springer, Heidelberg (2012). https://doi.org/10.1007/978-3-642-34478-7_57

17. Kim, M.-K., Kim, M., Oh, E., et al.: A review on the computational methods for emotional state estimation from the human EEG. Comput. Math. Methods Med. **2013**, 13 (2013)

18. Pankow, A., et al.: Altered amygdala activation in schizophrenia patients during emotion processing. Schizophr. Res. **150**(1), 101–106 (2013)

19. Zhang, D., Zhao, Y., Liu, Y., Tan, S.: Perception of the duration of emotional faces in schizophrenic patients. Sci. Rep. **6**, 22280 (2016)

Modeling an Augmented Reality Game Environment to Enhance Behavior of ADHD Patients

Saad Alqithami[✉], Musaad Alzahrani, Abdulkareem Alzahrani, and Ahmed Mostafa

Department of Computer Science, Albaha University, Albaha, Saudi Arabia
{salqithami,malzahr,ao.alzahrani,amyosof}@bu.edu.sa

Abstract. The paper generically models an augmented reality game-based environment to project the gamification of an online cognitive behavioral therapist that performs instant measurements for patients with a predefined Attention Deficit Hyperactivity Disorder (ADHD). ADHD is one of the most common neurodevelopmental disorders in which patients have difficulties related to inattention, hyperactivity, and impulsivity. Those patients are in need for a psychological therapy; the use of cognitive behavioral therapy as a firmly-established treatment is to help in enhancing the way they think and behave. A major limitation in traditional cognitive behavioral therapies is that therapists may face difficulty to optimize patients' neuropsychological stimulus following a specified treatment plan, i.e., therapists struggle to draw clear images when stimulating patients' mindset to a point where they should be. Other limitations recognized here include availability, accessibility and level-of-experience of the therapists. Therefore, the paper present a gamification model, we term as "*AR-Therapist*," in order to take advantages of augmented reality developments to engage patients in both real and virtual game-based environments. The model provides an on-time measurements of patients' progress throughout the treatment sessions which, in result, overcomes limitations observed in traditional cognitive behavioral therapies.

Keywords: Gamification · ADHD · Cognitive Behavioral Therapy · Augmented Reality

1 Introduction

Attention Deficit-Hyperactivity Disorder is an increasing concern in the past few decades. It has undefined etiology as a heterogeneous developmental disorder leading to bias and extensive diagnostic evaluations when examining patients through traditional clinical interviews and rating of patients' behaviors [1,3]. ADHD in underage patients can be observed in patients' hyperactivity and inability control their impulses and may have trouble paying attention which, in result, will intervene with their daily lives. In adulthood, patients with ADHD may have trouble managing time, being organized, setting goals, and holding down a job, which may lead to problems with relationships, self-esteem, and addiction. The treatment of many psychological disorders, such as ADHD,

© Springer Nature Switzerland AG 2019
P. Liang et al. (Eds.): BI 2019, LNAI 11976, pp. 179–188, 2019.
https://doi.org/10.1007/978-3-030-37078-7_18

can be through a well-known type of psychotherapy called *Cognitive Behavioral Therapy* (CBT). CBT involves patients in multiple psychosocial interventions in order to improve the status of their current mental health. This treatment requires patients to go through multiple sessions with specialized therapists. In ADHD, those sessions can be of increasing order of difficulty to help patients expand their cognitive capabilities to overcome current behavioral limitations.

ADHD has undefined etiology as a heterogeneous developmental disorder to involve hyperactivity and distractibility as well as difficulties with constant attention, impulsive control disorder and impaired cognitive flexibility, especially in problem solving and behavioral management [7]. Many studies have indicated the potential benefits of VR and AR exposure therapy for many types of mental disorders [4,6,16,18,21,22]. In a study by Parsons, et al. [21], attention performance was compared between 10 children with ADHD and 10 normal control children in a VR classroom. The results showed that children with ADHD are more impacted by distraction in the VR classroom. In spite of that, Ben-Moussa et al. [5] proposed a conceptual design of an exposure therapy system for patients with a social anxiety disorder. The proposed system integrates the AR and VR technologies through a simulated environment. It provides more effective exposure therapy solutions for patient with social anxiety disorder due to exploiting the benefits of AR and VR.

This highlights the importance of utilizing immersive technologies, e.g., Augmented Reality (AR) and Virtual Reality (VR), for their promising results that are stated in previous studies [4,6,16,18,21,22]. As the name suggests, AR-technology combines real and virtual contents, rendered in 3D, to be interactive in real-time [2]. Whereas, VR-technology replaces the real world with a computer-generated graphics via head mounted display [9]. In other words, the user in VR environment totally isolated from the real-world while the AR optimizes the interactions in the real world [8]. In AR, the environment is real but augmented with virtual objects from the system as it bridges the gap between real and virtual world in a seamless way [11]. The information presented by the immersed virtual objects enhances a user's perception of- and interaction with- the real-world [2]. The rendering have much lower requirements in AR than in VR because VR completely replace the real world with the virtual environment which makes display devices used in AR have less stringent requirements than that used in VR systems. Also, the tracking and sensing requirements for AR are much stricter than those for VR systems [2]. This can be of great benefits for ADHD patients because it enables therapists to collect more information about the behavior of the patient which helps in the diagnosis and treatment procedures.

It is unfortunate that proposed solutions fail to overcome language and cultural barriers for diverse patients and to employ the power of augmented reality by rendering 3D objects and avatars rather than solid textual instructions leading to increase in patients' engagements and speeding-up the recovery. The believe is that there is no statistically significant difference between the ADHD patients who will be treated using this system and those who are treated by traditional CBT; although, online CBT may exceed traditional ones methods by accelerating recovery time and saving money and resources for both government and patients. This is due to achieving a concept of "*a therapist for each patient*" as the system mimics the therapist roles through an augmented reality development that provide it with features including: adaptiveness, smartness, respon-siveness, and accuracy. Other advantages are availability, accessibility, and assurance of the therapist's level-of-experience which cannot be guaranteed in traditional CBT.

Therefore, this paper models a system using a game pipeline and propose measure-ments that can be used to test the validity of using an online system as CBT, which we called "*AR-Therapist*". The believe tackled here is that an increase in patient correct attention, e.g., when selecting a predefined object, contributes positively to their per-formance index which means they are following along with their treatment plan; the opposite should be true when they fail to achieve their assigned tasks. The result of this can be observed through applying the measurements we propose here on an augmented reality game.

2   Generic Design of the System

The high-level design, shown in Fig. 1, represents the main components of the AR-Therapist and the therapist-patient relationships within the online AR-environment.

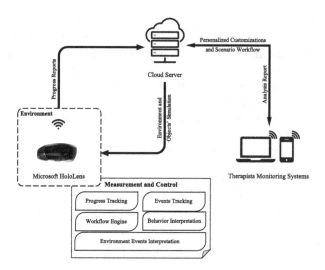

Fig. 1. The system architecture for AR-Therapist

Patients play-to-win the game that has been designed to attract their attention through the guidance of voice/sign instructions. Movement and actions are measured and then stored in a database to allow a therapist to monitor the treatment progress. Next, we present the logical flow presented in Fig. 1.

2.1   Formal Game Pipeline:

The AR-Therapist is based on the following 8-profiles for the generic assembly of the game pipeline.

1. **Treatment** is the whole treatment system (i.e., AR-Therapist). Its profile is a tuple of: ⟨Patient, Doctor, Game, {Treatment-Program}⟩
   - *Patient*: Each patient will have his/her own profile. The profile has to be complete for the patient to join the treatment program.
   - *Doctor*: Each doctor will have his/her own profile.
   - *Game*: The game has to be defined by the doctor considering patients current mental state and the disorder severity.
   - {*Treatment-Program*}: Players will go through a treatment program that includes playing an AR-based game consisting of a set of levels.
2. **Patient**: ⟨ID, Level, Performance-Index, {Preference}⟩
   - *ID*: is a short identification as a name or a referral number used by the doctor to define a patient.
   - *Level*: The level to where the patient has arrived in the treatment plan.
   - *Performance-Index*: The current value of performance the patient has achieved throughout the game.
   - {*Preference*}: The set of predefined preferences for a patient considering other psychological disorders that may affect current design and methodology of the treatment plan.
3. **Doctor**: ⟨ID, Experience, Involvement⟩
   - *ID*: The doctor has to have his/her own profile that is different from other therapists or psychological centers. This will give the doctor an access to the patient profiles and progress reports to allow for further evaluations.
   - *Experience*: Experience level of the doctor is useful in allowing access to more complex/detailed data of the patients.
   - *Involvement*: The level of engagement within the treatment process which allow the doctor to get involved in the game and in the reporting progress along the way of the patient assigned treatment.
4. **Game**: ⟨Type, {Level}⟩
   - *Type*: The type of the game to be played that has to be suitable for the patient. E.g., drag-and-drop and multiple-choices.
   - {*Level*}: The game consists of a set of levels that have different levels of complexity.
5. **Level**: ⟨{Object}, Max-Time, Effects⟩
   - {*Object*}: Maximum set of objects used in this level.
   - *Max-Time*: A predefined maximum time for the whole level to be completed or aport otherwise.
   - *Effects*: Simple directional voice or instructions used for guidance in case of a remote following.
6. **Object**: ⟨Shape, Size, Random-Location, Visibility⟩
   - *Shape*: The structure of an object has to be predefined beforehand the start of a session.
   - *Size*: The size of an object will depend on the location and closeness from the player focal point.
   - *Random-Location*: The initial distribution of objects around the real-environment.
   - *Visibility*: The appearance of one object after another.

7. **Treatment – Program**: $\langle\{$Game-session$\}$, Performance-Measures, Duration$\rangle$
   - *{Game-session}*: The set of game session to complete the treatment program.
   - *Performance-Measures*: The performance in one session reports the correct, incorrect and uncompleted tried the patient has gone through in one session.
   - *Duration*: The maximum treatment time for the whole treatment program. E.g., 20-min to complete the treatment.
8. **Game – session**: $\langle$Level, Timer, Current-location, Number-of-tries$\rangle$
   - *Level*: The game level has to be defined beforehand. The initial level is defined in the patient profile and player can move from one level to another asynchronously depending on his/her achievement in the session and then the patient profile is updated.
   - *Timer*: To count the response time for the patients.
   - *Current location*: To track current location of the patient for measuring closeness from objects within a session.
   - *Number of tries*: The repetition of tries within one session to include correct, incorrect and uncompleted tries. E.g., the number of collected target objects the patient has correctly collected in one session.

To this end, we have introduced Eight-profiles that when merged will best formulate the AR-Therapist model. Next, we show the process of combining those profiles into a conceptual model.

2.2 General Conceptual Model

The conceptual model consists of Four layers: (1) interface layer, (2) configuration layer, (3) run-time layer, and (4) storage layer. Figure 2 depicts an architecture of the AR-Therapist based on the profiles presented in previous section.

1. **The Interface Layer** contains the interfaces used for accessing other layers. For instance, doctor can use a user-friendly interface to configure the treatment plan, whereas player interact with the game through Augmented Reality glasses.
2. **The Configuration Layer** consists of the "treatment plan configuration" component, where the doctor can add new treatment plan and configure the existence ones.
3. **The Run-Time Layer** has the components that interact with the player while the game is running. These components are: the Context Agent component and the Game component.
   - **The Context-Agent Component** retrieves the player treatment plan, and his/her performance in order to controlling the current game-session and guiding the player based on the treatment plan. Furthermore, the context agent capable of monitoring all the player's behaviors and interactions with the environment, logging "Ethically" the needed data, and calculating the player's performance following the measurements highlighted in the next section. Thus, the agent gains a deeper understanding of the patient to enrich the AR-Therapist model based on appropriate reasoning techniques. This in turns can be utilized in the future for suggesting the most optimal treatment plan for new patient (e.g. when facing a "cold start issue"). In addition, logging the needed data and reasoning them will

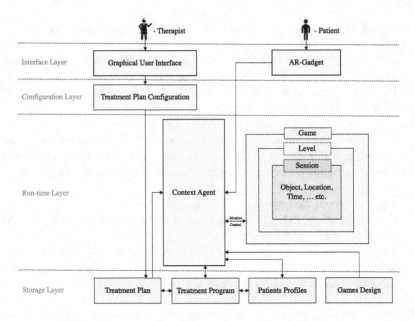

Fig. 2. The general conceptual model for AR-Therapist

reveal some of hidden information that can be valuable for advancing research on treating patients with ADHD.

On the other hand, there is a persistent need for employing intelligent agent capable of performing the tasks explained above since the AR-Therapist model is used by the patients with the aid of their families and under the doctors' remote supervision (i.e. ambient assistant living "AAL"). A prior use of AAL is observed in the literature for fall detection [10], and for monitoring elderly in their homes using video surveillance [15].

- **The Game Component** involves levels of game; each level has its own games-sessions, and each game-session in turns contains the appropriate difficulty level that characterizes the maximum game-session time, used objects, and their locations.

4. **The Storage Layer** involves components that store data about the treatment-plan, treatment-program, player profile, and game levels and objects. Each component follows an appropriate common specification for structuring the related data. Thus, the system's extensibility can be guaranteed as well as the integration with the existence E-Health systems.

3 Performance Measures

One of the most important issues in diagnosing and treatment of ADHD is to determine a set of accurate performance measures. These performance measures should have the ability to differentiate accurately between patients having ADHD symptoms and others who do not have it. According to psychiatric recommendations, it would be better to collect these performance measures from patients within different environments such as at homes or schools [19]. There are many tests used to diagnose patients with ADHD [17]. Continuous Performance Tests (CPT) are the most popular laboratory-based test supporting clinical diagnosis [13,14,25]. CPT is usually a computer-based test that aims to measure attention and impulsivity. CPT involves the individual and random presentation of a series of visual or auditory stimuli that changes rapidly over a period of time. Patients are informed to respond to the "target" stimulus and avoiding a "non-target" stimulus. The test provides summary statistics of performance parameters (e.g., response time, average response time, response time standard deviation, omission errors, and commission errors). These parameters have been shown to be useful in the detection of ADHD [23].

An important limitation of traditional CPTs is low ecological validity [24]. Ecological validity means the degree to which a performance test produces results similar to those produced in real life [20]. One approach to improve assessment methods which offers better ecological validity is CPTs based on VR, such as the Aula Nesplora test [12]. Those approaches have an advantage of being more realistic and ecologically-valid environment while still having the ability to assess the degree of ADHD severity. Using AR instead of VR will further improve the ecological validity of the performed test. Here, multiple performance measures is used to provide ADHD diagnostics and treatment assessment. Thus, let us assume that we have an AR-game that frequently present an interactive environment with the following assumptions:

T: Total number of tries progressed in one session, which is a combination of:
   a. Number of correct tries in one session, i.e., C,
   b. Number of incorrect tries (due to omission or commission errors) in one session, i.e., I, and
   c. Number of uncompleted tries in one session, i.e., K.

Then, the performance measures that will be used for providing ADHD diagnostics and treatment assessment include:

- **Correct Response Times** (*CRT*): The percentage of measuring attention deficits for the time spent on the correct tries. CRT measures the time period that a patient takes to make a correct try (i.e., choose the target object in the try). The longer the CRT is, the more likely the patient has attention deficit and less time to focus on tasks. As a result, the patient takes longer time comparing to normal subjects when performing a task (i.e., choosing the target/correct object in our case).
- **Mean of the CRT** (*M*): To compare with correct response time to make sure they follow opposite relation to one another, i.e., $M = \sum_{i=1}^{C} CRT_i / C$.
- **Standard deviation of the CRT** (*SD*): Indicative of impulsive and hyperactive symptoms of the patient. The higher the value of SD, the more probability that patient suffers from impulsivity/hyperactivity and difficulty in controlling his/her moves after a certain period of time. As a result, the patient starts periodically to move with no destination. I.e., $SD = \sqrt{\sum_{i=1}^{C} (CRT_i - M)^2 / \{C - 1\}}$.

- **Try time** (θ): The maximum allowed time to complete one try within a session.
- **Omission errors** (OE): The absence of any response during a try period to be used to measure inattention.
- **Commission errors** (CE): The response to non-target stimuli which to be used to measure impulsivity.
- **Engagement Factor** (GF): It indicates the patient engagement level with the game. In a case, a patient is considered to be engaged in the game if s/he continues to play the game. In contrast, the patient is considered to be not engaged if s/he stops the game before completing all tries in a session. Thus, GF is defined as the number of correct and incorrect tries divided by the total number of tries in the session, i.e., $GF = \{C+I\}/T$.
- **Inattention Factor** (IAF): It represents the percentage of patient's inattention, i.e., $IAF = OE/\{C+I\}$. In a case, the inattention of a patient increases when s/he makes Omission Errors (OE) in a session, i.e., when the patient does not choose any of the objects appearing to him/her. The number of uncompleted tries (K) in a session should be excluded when indicating IAF, i.e., $C+I = T-K$.
- **Impulsivity Factor** (IMF): Indicates the percentage of patient's impulsivity observed in his/her behavior within a session. IMF is defined as the number commission errors divided by number of correct and incorrect tries, i.e., $IMF = CE/\{C+I\}$. We also excluded K when defining IMF. In our case, a patient who suffers from impulsivity will likely make more commission errors because impulsivity will prevent him/her from focusing when choosing an object.
- **Error Factor** (EF): It represents the percentage of the error rate during a session, i.e., $EF = \{OE+CE\}/\{C+I\}$. The error in the case includes omission and commission errors excluding K. Thus, $EF = IAF + IMF$.
- **Correct Response Factor** (CRF): A percentage of total correct response-time relative to maximum allowed time for all correct tries, i.e., $CRF = \{\sum_{i=1}^{c} CRT_i\}/\{C \times \theta\}$. In a case, CRF should be negatively affected by the amount of time that the patient takes when s/he makes an incorrect try. Thus, we define CRF as the total summation of CRTs to the actual time of the game during the session (GT). In this case, CRF will be 100% if the patient makes all tries correctly. Otherwise, it will decrease depending on the total amount of time spent by the patient on I.
- **Performance Index** (PI): It reflects the single measure for the overall performance of the patient which depends on the correct response, error, and engagement factors.

$$PI = \left[\frac{(1-CRF)+(1-EF)}{2} \right] \times GF$$

PI is a composite score which measures the overall performance of a patient. In a case, the PI is affected positively by the patients' CRF and negatively by the patients' EF. In addition, we need to take into account different possible scenarios that can happen in the game session. One possible scenario is that the patient does not complete all the tries in the session. The patient can make one correct try and stop the game before finishing all the tries in the session. If we only considered the CRF and EF, the PI in the case would be the highest. In order to prevent this from happening, we consider the GF in the definition of PI. Another possible scenario is that we have two patients who have the same CRF, EF, and GF but different GT. In this case, they will have the same PI. However, the patient who has less GT should have a higher PI. Thus, we considered the ratio of GT to the maximum session time when defining the PI.

4   Conclusion and Future Work

The paper proposed a gamification system, called "AR-Therapist", as an online cognitive behavioral therapist to help in treating patients with a predefined ADHD symptoms. The purpose is to replace traditional CBT with a more advanced virtual one that may exceed traditional CBT methods by accelerating recovery time and saving money and resources. AR-Therapist achieves an excellent accessibility level to every patient in need, as it mimics the therapist roles through an augmented reality environment that provides it with features including: adaptiveness, smartness, responsiveness, and accuracy. Other advantages are availability and assurance of the therapist's level-of-experience which cannot be guaranteed in traditional CBT. We are working to implement a simulated augmented reality environment using a simple game, where the results should lead to an increase in patient correct attention (e.g., choosing a predefined object) which contributes positively to their performance index, i.e., they are following along with their treatment plan; the opposite is correct when they fail to achieve their assigned tasks. Future research should provide extensive empirical case studies that will support this area of research with a rich and valuable data when collected ethically and will introduce interesting findings when analyzed.

Acknowledgement. This work was funded by the Deanship of Scientific Research at Albaha University, Saudi Arabia [grant number: 1439/4]. Any opinions, findings, and conclusions or recommendations expressed in this material are those of the author(s) and do not necessarily reflect the views of the Deanship of Scientific Research or Albaha University.

References

1. Abikoff, H., Courtney, M., Pelham, W.E., Koplewicz, H.S.: Teachers' ratings of disruptive behaviors: the influence of halo effects. J. Abnormal Child Psychol. **21**(5), 519–533 (1993)
2. Azuma, R.T.: A survey of augmented reality. Presence: Teleoperators Virtual Environ. **6**(4), 355–385 (1997)
3. Barkley, R.A.: The ecological validity of laboratory and analogue assessment methods of ADHD symptoms. J. Abnormal Child Psychol. **19**(2), 149–178 (1991)
4. Beard, L., Wilson, K., Morra, D., Keelan, J.: A survey of health-related activities on second life. J. Med. Internet Res. **11**(2), e17 (2009)
5. Ben-Moussa, M., Rubo, M., Debracque, C., Lange, W.G.: DJINNI: a novel technology supported exposure therapy paradigm for SAD combining virtual reality and augmented reality. Front. Psychiatry **8**, 26 (2017)
6. Bickmore, T.W., Mitchell, S.E., Jack, B.W., Paasche-Orlow, M.K., Pfeifer, L.M., O'Donnell, J.: Response to a relational agent by hospital patients with depressive symptoms. Interact. Comput. **22**(4), 289–298 (2010)
7. Biederman, J.: Attention-deficit/hyperactivity disorder: a selective overview. Biol. Psychiatry **57**(11), 1215–1220 (2005)

8. Billinghurst, M., Clark, A., Lee, G.: A survey of augmented reality. Found. Trends® Hum.-Comput. Interact. **8**(2–3), 73–272 (2015)
9. Burdea, G.C., Coiffet, P.: Virtual Reality Technology. Wiley, Hoboken (2003)
10. Chan, E., Wang, D., Pasquier, M.: Towards intelligent self-care: multi-sensor monitoring and neuro-fuzzy behavior modelling. In: 2008 IEEE International Conference on Systems, Man and Cybernetics, pp. 3083–3088, October 2008
11. Chang, G., Morreale, P., Medicherla, P.: Applications of augmented reality systems in education. In: Society for Information Technology & Teacher Education International Conference, pp. 1380–1385. Association for the Advancement of Computing in Education (AACE) (2010)
12. Climent, G., Banterla, F., Iriarte, Y.: Aula: Theoretical Manual. Nesplora, San Sebastian (2011)
13. Conners, C.K., Staff, M., Connelly, V., Campbell, S., MacLean, M., Barnes, J.: Conners' continuous performance Test II (CPT II v. 5). Multi-Health Syst. Inc. **29**, 175–196 (2000)
14. Edwards, M.C., Gardner, E.S., Chelonis, J.J., Schulz, E.G., Flake, R.A., Diaz, P.F.: Estimates of the validity and utility of the conners' continuous performance test in the assessment of inattentive and/or hyperactive-impulsive behaviors in children. J. Abnormal Child Psychol. **35**(3), 393–404 (2007)
15. Foroughi, H., Aski, B.S., Pourreza, H.: Intelligent video surveillance for monitoring fall detection of elderly in home environments. In: 2008 11th International Conference on Computer and Information Technology, pp. 219–224, December 2008
16. Gorini, A., Gaggioli, A., Vigna, C., Riva, G.: A second life for ehealth: prospects for the use of 3-D virtual worlds in clinical psychology. J. Med. Internet Res. **10**(3), e21 (2008)
17. Hall, C.L., et al.: The clinical utility of the continuous performance test and objective measures of activity for diagnosing and monitoring ADHD in children: a systematic review. Eur. Child Adolesc. Psychiatry **25**(7), 677–699 (2016)
18. Meyerbröker, K., Emmelkamp, P.M.: Virtual reality exposure therapy in anxiety disorders: a systematic review of process-and-outcome studies. Depression Anxiety **27**(10), 933–944 (2010)
19. Morales-Hidalgo, P., Hernández-Martínez, C., Vera, M., Voltas, N., Canals, J.: Psychometric properties of the conners-3 and conners early childhood indexes in a Spanish school population. Int. J. Clin. Health Psychol. **17**(1), 85–96 (2017)
20. Negut, A., Jurma, A.M., David, D.: Virtual-reality-based attention assessment of ADHD: ClinicaVR: Classroom-CPT versus a traditional continuous performance test. Child Neuropsychol. **23**(6), 692–712 (2017)
21. Parsons, T.D., Bowerly, T., Buckwalter, J.G., Rizzo, A.A.: A controlled clinical comparison of attention performance in children with ADHD in a virtual reality classroom compared to standard neuropsychological methods. Child Neuropsychol. **13**(4), 363–381 (2007)
22. Parsons, T.D., Rizzo, A.A., Rogers, S., York, P.: Virtual reality in paediatric rehabilitation: a review. Dev. Neurorehabilitation **12**(4), 224–238 (2009)
23. Rapport, M.D., Chung, K.M., Shore, G., Denney, C.B., Isaacs, P.: Upgrading the science and technology of assessment and diagnosis: laboratory and clinic-based assessment of children with ADHD. J. Clin. Child Psychol. **29**(4), 555–568 (2000)
24. Rodríguez, C., Areces, D., García, T., Cueli, M., González-Castro, P.: Comparison between two continuous performance tests for identifying ADHD: traditional vs. virtual reality. Int. J. Clin. Health Psychol. **18**(3), 254–263 (2018)
25. Vogt, C., Williams, T.: Early identification of stimulant treatment responders, partial responders and non-responders using objective measures in children and adolescents with hyperkinetic disorder. Child Adolesc. Mental Health **16**(3), 144–149 (2011)

Exploring the Characterization and Classification of EEG Signals for a Computer-Aided Epilepsy Diagnosis System

Emil Vega-Gualán[1,3(✉)], Andrés Vargas[3(✉)], Miguel Becerra[2,3(✉)], Ana Umaquinga[3,5], Jaime A. Riascos[3,4(✉)], and Diego Peluffo[3,4]

[1] Yachay Tech University, Urcuqui, Ecuador
emil.vega@yachaytech.edu.ec
[2] Institución Universitaria Pascual Bravo, Medellín, Colombia
migb2b@gmail.com
[3] SDAS Research Group, Urcuqui, Ecuador
andres.vargas.innovatec@gmail.com
[4] Corporación Universitaria Autónoma de Nariño, Pasto, Colombia
jarsalas@inf.ufrgs.br, diegohpo@gmail.com
[5] Universidad Técnica del Norte, Ibarra, Ecuador
https://www.yachaytech.edu.ec/
http://www.sdas-group.com

Abstract. Epilepsy occurs when localized electrical activity of neurons suffer from an imbalance. One of the most adequate methods for diagnosing and monitoring is via the analysis of electroencephalographic (EEG) signals. Despite there is a wide range of alternatives to characterize and classify EEG signals for epilepsy analysis purposes, many key aspects related to accuracy and physiological interpretation are still considered as open issues. In this paper, this work performs an exploratory study in order to identify the most adequate frequently-used methods for characterizing and classifying epileptic seizures. In this regard, a comparative study is carried out on several subsets of features using four representative classifiers: Linear Discriminant Analysis (LDA), Quadratic Discriminant Analysis (QDA), K-Nearest Neighbor (KNN), and Support Vector Machine (SVM). The framework uses a well-known epilepsy dataset and runs several experiments for two and three classification problems. The results suggest that DWT decomposition with SVM is the most suitable combination.

Keywords: Electroencephalogram (EEG) · Epilepsy diagnosis · K-Nearest Neighbors (KNN) · Linear Discriminant Analysis (LDA) · Quadratic Discriminant Analysis (QDA) · Support Vector Machine (SVM)

© Springer Nature Switzerland AG 2019
P. Liang et al. (Eds.): BI 2019, LNAI 11976, pp. 189–198, 2019.
https://doi.org/10.1007/978-3-030-37078-7_19

1   Introduction

Epilepsy has become the third most common neurological disorder after stroke and dementia. According to the World Health Organization (WHO), epilepsy affects 0.5–1.5% of the world population, mainly children under 10 and people over 65, and it is more common in developing countries and disadvantaged socioeconomic classes [2]. The seizures are the hallmark for an epilepsy diagnosis. They are recurrent but infrequently and unprovoked signals caused by the synchronized electrical discharge of a large number of neurons [9]. Since epilepsy occurs when the localized electrical activity of neurons suffers from an imbalance, analyzing the electroencephalographic signals (EEG) is one of the most suitable methods to diagnosis this disorder.

Most of the computer-aided systems for diagnosis epilepsy use EEG because it allows rapid and visual inspection of seizures, not only when they are occurring, but also the pre-occurrence- and between- seizures [12]. The common procedure for developing automatic diagnostic-assistance systems based on EEG has five stages (citation): EEG signal acquisition, pre-processing, signal characterization, classification and in-context interpretation (visualization). Zhou and colleagues [13] developed an epileptic seizure detection using the raw EEG Signals (temporal approach), meanwhile, Tsipouras [10] studies the epilepsy classification using spectral information of EEG signals.

This work aims to contribute with an exploratory study about the feature extraction and classification techniques on seizure detection. The proposed framework includes the typical stages described above as follows: a simple amplitude normalization for pre-processing signals. Subsequently, features are extracted using statistical measures on both the original signals and spectral transformation thereof (Discrete Wavelet Transformation, DWT). Afterward, a set of features is chosen by applying feature selection methods such as `Bestfirst` and `Ranker`. Then, the selected features are classified by using four of the most representative classification approaches for EEG: Linear Discriminant Analysis (LDA), Quadratic Discriminant Analysis (QDA), K-Nearest Neighbor (KNN) and Support Vector Machine (SVM). The framework uses the "Epileptic Seizure Recognition Data Set"[1] for running several tests in order to explore as much as possible the proposed classifiers and features. The outcome of the study points out the DWT features and SVM classifier as the most suitable combination.

2   Materials and Methods

2.1   Dataset

The "Epileptic Seizure Recognition Data Set" is composed of 500 individuals with 4097 data points of 23.6 seconds each one. This dataset is later divided and shuffled every 4097 data points into 23 chunks which contain 178 data points for 1-second [8]. Therefore, it contains a matrix of dimension 11500×178. The last column represents the labels (1,2,3,4,5) as follows:

---

[1] Available on UCI machine learning repository https://archive.ics.uci.edu/ml.

1. Seizure activity
2. EEG signal from the area where the tumor is
3. EEG activity from the healthy brain area
4. EEG signal from eyes closed
5. EEG signal from eyes open.

2.2 EEG Pipeline

Pre-processing: The signals are normalized in order to remove offset levels using the Eq. 1.

$$S = \frac{S - \bar{S}}{max\,|S|} \tag{1}$$

where S is the signal, $max\,|S|$ is the maximum absolute value of the signal, and $\bar{S}$ is the mean of the signal.

Signal Decomposition: Discrete Wavelet Transform (DWT) decomposes a signal recurrently into two sub-signals (approach and detail) with less resolution regarding the frequency, known as coefficients [6]. Daubechies family is used in this work (order four) through the MATLAB function `wavedec`.

Characterization: Considering previous works of EEG signal [1,11], several features are used as follows:

Number	Type	Description
$x^{(1)} \dots x^{(15)}$	Temporal	Statistical features
		Entropy
$x^{(16)} \dots x^{(27)}$	Morphological	Area under the curve
		Amplitude change
		Energy
$x^{(28)} \dots x^{(35)}$	Spectral	Fourier transform
		(Best features)
$x^{(36)} \dots x^{(221)}$	Representative	DWT
		(Temporal and morphological features)

This results in a feature matrix of dimension 11500×221. Where 221 is the total number of features normalized through the Eq. 1.

Feature Selection: In order to use the most significant features, two methods from Weka [5] are used to create a smaller subset:

(i) Using the `CfsSubsetEval` as attribute evaluator and `BestFirst` as search method [4]. This is applied to the whole feature matrix.

(ii) Using the `InfoGainAttributeEval` as attribute evaluator and `Ranker` as search method. This is applied to the outcome of the previous step.

Obtaining thus a final feature matrix of size 11500×9. Where 9 is the total number of features used in this work. Most of the features coming from the DWT subset.

2.3   Classification

These features are used to train and evaluate four different classifiers. In this sense, a representative method of each typology (Distance-based (k-NN), model-based (LDA, QDA) and SVM (data-driven)) are used [3, 7]:

– **Linear Discriminant Analysis (LDA):** LDA creates a hyperplane by the projection of the co-variance matrices that separates the classes and estimates the probability that a new data belongs to each class (maximum probability).
– **Quadratic Discriminant Analysis (QDA):** It is a variant of LDA, where an individual co-variance matrix is estimated for every class of observations.
– **K-Nearest Neighbour (k-NN):** k-NN assigns the classification label following the largest posterior probability among to nearest neighbor's values, which is calculated using a metric distance.
– **Support Vector Machine (SVM):** SVM finds the optimal hyperplane (in a N-dimensional space) that separates the data by maximizing the margin between the classes.

For all the experiments (see the next section), the classifiers were trained with 10 iterations using 80% of the data and the rest for testing. All experiments were run in Matlab using the classifier settings shown in Table 1. Finally, Fig. 1 summarizes graphically the methodology used in this research.

Table 1. Classifiers settings

Classifier	Settings
LDC	No regularization
	All dimensions
QDC	No regularization
	All dimensions
KNN	k is optimized with respect to the leave-one-out error on the dataset
SVM	Kernel function: Gaussian
	Kernel scale: 3
	Box constraint: 1
	Standardize: true
	Specify the class names

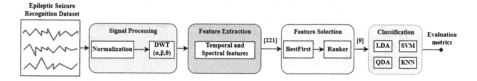

Fig. 1. Block diagram of the proposed methodology.

3   Results and Discussion

This work uses the Dunn test (Kruskal-Wallis with bonferroni correction) for performing comparisons among the classifiers. In order to evaluate as much as possible the performance of the classifiers, the final matrix with the selected features is used for doing five tests. All experiments are performed regarding the target class 1 which is the most important class for this research (seizure activity). Additionally, for evaluating the capacity of the methodology, the five classes of the original dataset were restructured to work with binary and three-class cases instead of multiclass cases (more complex and challenging). The combination of classes are described as follows:

3.1   Experiment 1

In this experiment the classes 2, 3, 4 and 5 are merged into a single class, with class 1 as target for classification. As result, there are significant differences (chi-squared = 30.8512, df = 3) among both LDA and QDA with KNN and SVM. SVM achieves the lowest error rate followed nearly by kNN. Figure 2 shows the result.

3.2   Experiment 2

For this experiment, we tried to classify the region where the seizure is presented; therefore, the seizure (class 1) and tumor localization (class 2) are used alone, and the rest of the classes are removed from the dataset. We found that LDA is significant different with SVM, meanwhile QDA is with both SVM and KNN (chi-squared = 34.929, df = 3). SVM presents again the lower error rate followed nearly by KNN. The worst of them is QDC. Figure 3 shows the result.

3.3   Experiment 3

Here, we intended to distinguished the seizure activity and the area where the tumor localization is, so the classes 1 and 2 form two classes individually, and the rest of the classes as a single one. This experiment presents the similar results as the previous one, that is, QDA as the worst classifier with a significant difference with both SVM and KNN, meanwhile LDA only with SVM. SVM achieves again the lowest error rate. Figure 4 shows the result.

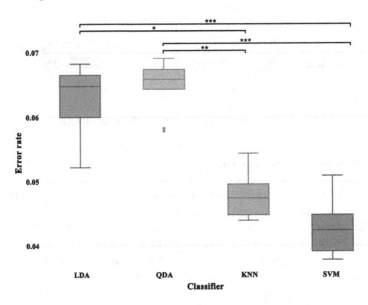

Fig. 2. Experiment 1 - Comparison of classifiers. Significant difference p-value $* <= 0.05, ** <= 0.001, *** <= 0.0001$.

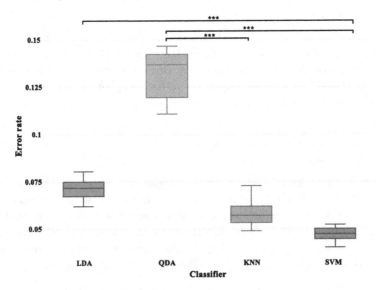

Fig. 3. Experiment 2 - Comparison of classifiers. Significant difference p-value $* <= 0.05, ** <= 0.001, *** <= 0.0001$.

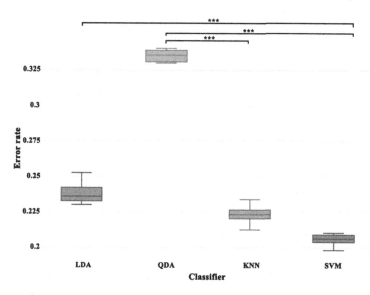

Fig. 4. Experiment 3 - Comparison of classifiers. Significant difference p-value $* <= 0.05, ** <= 0.001, * * * <= 0.0001$.

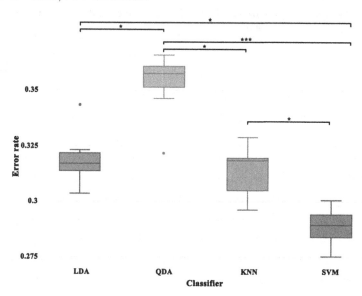

Fig. 5. Experiment 4 - Comparison of classifiers. Significant difference p-value $* <= 0.05, ** <= 0.001, * * * <= 0.0001$.

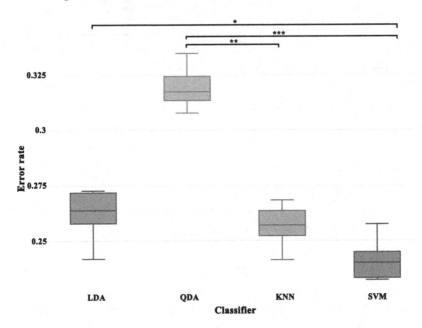

Fig. 6. Experiment 5 - Comparison of classifiers. Significant difference p-value $* <= 0.05, ** <= 0.001, *** <= 0.0001$.

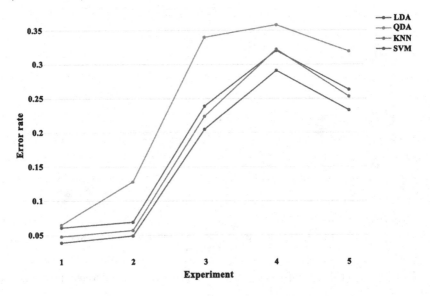

Fig. 7. Comparison of classifiers throughout their accuracies.

3.4   Experiment 4

In this opportunity the classes 1, 2 and 3 represent different classes individually, with the rest of classes removed from the dataset. Therefore, class 3 can distinguish the healthy brain area. The results suggest that SVM achieves the lowest error rate and significantly differs with the other classifiers (chi-squared = 31.7648, df = 3). Meanwhile, QDA obtains the higher error rate. Figure 5 shows the result.

3.5 Experiment 5

Continuing with the multiclass problem, the classes 1 and 2 were merged as a single class, meanwhile class 3 individually and classes 4 and 5 as a single one as well. The results present again SVM as the best classifiers. There are significant differences between LDA and SVM, and QDA with both SVM and KNN. Again the worst one is QDA. Figure 6 shows the result.

Finally, Fig. 7 presents the comparison among classifiers along the five experiments, evidencing the SVM's behavior as the best classifier.

4 Conclusions and Future Work

The comparison of the explored techniques suggests that the features from DWT decomposition along with Support Vector Machine are the best alternatives to build an Epilepsy-driven EEG analysis computer-aided system. The aim of combining several classes is to study the performance of classifiers, demonstrating that for binary classification both KNN and·SVM are good alternatives, meanwhile, for the three-class problem, QDA shows the worst performance and SVM the best. Indeed, the seizure data are hardly separable classes, therefore, a data-driven method like SVM with kernel solution was necessary. Other alternatives should be considered for future studies, among them the feature extraction and deep learning approaches. Besides, given the number of methods (dozens of methods), it is necessary to scale the study with more performance measures to be able to do a comparison with other studies.

Acknowledgments. Authors thank to the SDAS Research Group (www.sdas-group.com) for its valuable support.

References

1. Bajaj, V., Pachori, R.B.: EEG signal classification using empirical mode decomposition and support vector machine. In: Deep, K., Nagar, A., Pant, M., Bansal, J.C. (eds.) Proceedings of the International Conference on Soft Computing for Problem Solving (SocProS 2011) December 20-22, 2011. AISC, vol. 131, pp. 623–635. Springer, New Delhi (2012). https://doi.org/10.1007/978-81-322-0491-6_57
2. Christensen, J., Sidenius, P.: Epidemiology of epilepsy in adults: implementing the ILAE classification and terminology into population-based epidemiologic studies. Epilepsia **53**, 14–17 (2012)
3. Duda, R.O., Hart, P.E., Stork, D.G.: Pattern Classification, 2nd edn. Wiley, New York (2000)

4. Hall, M.A.: Correlation-based feature subset selection for machine learning. Ph.D. thesis, University of Waikato, Hamilton, New Zealand (1999)
5. Hall, M., Frank, E., Holmes, G., Pfahringer, B., Reutemann, P., Witten, I.H.: The WEKA data mining software: an update. SIGKDD Explor. 11(1), 10–18 (2009)
6. Ocak, H.: Automatic detection of epileptic seizures in EEG using discrete wavelet transform and approximate entropy. Expert Systems with Applications 36(2), 2027–2036 (2009)
7. Qazi, K.I., Lam, H., Xiao, B., Ouyang, G., Yin, X.: Classification of epilepsy using computational intelligence techniques. CAAI Trans. Intell. Technol. 1(2), 137 – 149 (2016). https://doi.org/10.1016/j.trit.2016.08.001, http://www.sciencedirect.com/science/article/pii/S2468232216300142
8. Qiuyi, W., Ernest, F.: Epileptic Seizure Recognition Data Set. UCI Machine learning repository (2017). https://archive.ics.uci.edu/ml/datasets/Epileptic+Seizure+Recognition
9. Stafstrom, C.E., Carmant, L.: Seizures and epilepsy: an overview for neuroscientists. Cold Spring Harbor Perspect. Med. 5(6), a022426 (2015). https://doi.org/10.1101/cshperspect.a022426
10. Tsipouras, M.G.: Spectral information of EEG signals with respect to epilepsy classification. EURASIP J. Adv. Sig. Process. 2019(1), 10 (2019). https://doi.org/10.1186/s13634-019-0606-8
11. Übeyli, E.D.: Statistics over features: EEG signals analysis. Comput. Biol. Med. 39(8), 733–741 (2009)
12. Wang, L., et al.: Automatic Epileptic Seizure Detection in EEG Signals Using Multi-Domain Feature Extraction and Nonlinear Analysis, May 2017. https://www.mdpi.com/1099-4300/19/6/222
13. Zhou, M., et al.: Epileptic seizure detection based on EEG signals and CNN. Front. Neuroinformatics 12, 95–95 (2018). https://doi.org/10.3389/fninf.2018.00095, https://www.ncbi.nlm.nih.gov/pubmed/30618700, 30618700[pmid]

Brain-Machine Intelligence and Brain-Inspired Computing

Evaluation of Identity Information Loss in EEG-Based Biometric Systems

Meriem Romaissa Boubakeur[1(✉)], Guoyin Wang[1], Ke Liu[1], and Karima Benatchba[2]

[1] Chongqing Key Laboratory of Computational Intelligence, Chongqing University of Posts and Telecommunications, Chongqing 400065, China
am_boubakeur@esi.dz
[2] Laboratoire des Méthodes de Conception de Systèmes, Ecole nationale Supérieure d'Informatique, 16270 Alger, Algeria

Abstract. Recently, electroencephalogram (EEG) has been used as a biometric modality. An EEG-based biometric system allows an automatic recognition of people based on their EEG signals. The quantity and quality of identity information extracted from EEG determine the performance of the EEG-based biometric system. In this paper, we evaluate the loss in identity information through different signal segmentation scenarios using Autoregressive model and K-Nearest Neighbor classifier. Our objective is to find some criteria linked to data segmentation allowing to reduce as far as possible the simulated loss of identity information. Experiments were conducted on EEG publicly available datasets collected in resting state for both opened and closed eyes. Results show that overlapped segmentation with longer segments' length stands the best to the simulated loss favoring larger percentages of overlap.

Keywords: EEG · Biometrics · Person identification · Segmentation

1 Introduction

EEG (electroencephalogram) is the electrical recording and measurement of brain activity on the surface of the human skull using electrodes. This measurement reflects the summation of small electrical impulses emitted by the huge amount of brain's neurons [2]. Recently, there is a growing interest towards using EEG in biometrics. EEG offers the advantage to be more secured compared to the other biometric traits. In fact, EEG cannot be captured at a distance since it is not exposed like iris or face. Moreover, intruder cannot force user to provide EEG because under-stress brain activity changes. One of the first investigations of EEG as a biometric trait was proposed in [14]. In this EEG-based biometric system, with Autoregressive (AR) model and Learning

This work has been supported by the National Natural Science Foundation of China under grant 61572091.

Vector Quantizer Network, Poulos et al. built a person identification method to distinguish between 4 subjects, who were in resting state when EEG was collected and could achieve correct classification scores at the rage of 72% to 81%. Paranjape et al. used Autoregressive models of various orders with discriminant functions and could distinguish between 40 subjects with an accuracy of 80% when 50% of data are used for tests [12]. Parisa et al. proposed the combination of AR model and competitive network to distinguish between 10 subjects using EEG collected in resting state. They could achieve correct classification scores at the range of 80% to 100% [11]. Rocca et al. could identify 45 subjects with high recognition rate (up to 98.73%) using Autoregressive stochastic modeling and polynomial regression-based classification where EEG were collected in resting state [8]. Besides, more reviews about EEG-based biometric systems can be found in [1–4]. In EEG-based biometrics systems, the most critical step is how to successfully obtain sufficient identity information from EEG signals for clear distinctions between subjects. However, obtaining the complete identity information from EEG signals is much more difficult than any other biometric traits. Several factors affect the quality and quantity of identity information for EEG-based biometric system:

1. It is hard to understand EEG signals since they reflect the whole brain activities (including background and task-related activities). Additionally, EEG signals are non-stationary.
2. The environment of collecting EEG signals affects their quality and might cause some damages in the identity information contained in EEG.
3. EEG consists of a set of multi-channel signals broadcasted at the same time. Channel selection determines the source of information handled by EEG-based biometric system. This selection might ignore some channels that contain a portion of the identity information.

Therefore, probable loss in quantity and quality of identity information arises from the uncertainty in its location. The evaluation of this loss through different data processing techniques allows to define criteria of how to locate the identity information as complete as possible in EEG-based biometric systems. In this paper, we evaluate the loss related to the quality of identity information through different data segmentation scenarios. Our objective is to illustrate the impact of this loss on the performance of EEG-based biometric systems. Additionally, we discuss some criteria linked to data segmentation allowing to reduce as far as possible the simulated loss of identity information. The remainder of this paper is organized as follows. In Sect. 2, the proposed methodology is presented. In Sect. 3 are reported and discussed the experimental results using EEG in resting state from Motor Movement/Imaginary datasets. Finally, in Sect. 4 conclusion and future works are summarized.

2   Methodology

In this section, we first present the architecture of EEG-based biometric system used. Then, we introduce our methodology in simulating and evaluating the loss in identity information. Finally, we describe our EEG-based biometric system.

2.1   Architecture of EEG-Based Biometric System

EEG-based biometric system includes two types of biometric applications namely identification and authentication. It is useful to distinguish between these two types since they have impact on the choice of performance evaluation of EEG-based biometric system (see Fig. 1) [13]. Identification can be either open-set identification or close-set identification. In this paper, we choose the architecture of close-set identification where any subject presented to the system for recognition is known to be enrolled in the system. In this case, we model the system as a multi-classification problem. Every subject is represented by a class label grouping his/ her EEG signals. Thereby, the system aims to find the appropriate class label (identity of subject) from a given EEG signal. For system evaluation, Cumulative Match Characteristic (CMC) curve is mostly recommended for close-set identification (identification rate for rank-k) [5,13]. In this paper, we use the classification accuracy which is the identification rate for rank-1 [9].

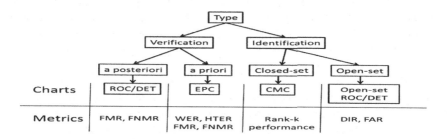

Fig. 1. Classification of different biometric performance charts [13].

2.2   Evaluation of Identity Information Loss in EEG-Based Biometric System

Our contribution illustrates the impact of identity information loss on the performance of EEG-based biometric system through two main steps:

Simulation of Identity Information Loss. The first step involves the simulation of identity information loss by affecting purposely the quality of EEG signals. The deterioration in EEG signal's quality is made progressively to produce different level of EEG signals' quality leading consequently to different level of identity information quality. One of signal processing techniques that might cause some loss of information is downsampling. Hence, the process implies in resampling the original EEG signal with different sampling frequencies $S_1 = \{f_1, f_2, ..., f_n\}$ where:

- f_n is the original sampling frequency and $f_i < f_n$ for $1 \leq i \leq n - 1$.
- $EEG_{S_1} = \{EEG_{f_1}, EEG_{f_2}, ..., EEG_{f_n}\}$, $EEG_{f_i} = resample\,(EEG_{f_n}, f_i)$ and EEG_{f_n} is the original EEG signal.

Localization of Identity Information by Segmentation. The second step attempts to locate the identity information by splitting EEG signal into segments (sweeps). We propose to consider two types of segmentation namely:

Disjoint Segmentation: EEG signal of duration D is divided into segments of duration d, where the number of data points per the corresponding segment $l = d * sampling frequency$, and the number of data points per EEG signal $L = D * sampling frequency$. In this case of segmentation, the end of one segment is the beginning of the next one. The total number of segments gotten from EEG signal $N = \left[\frac{D}{d}\right] = \left[\frac{L}{l}\right]$.

Overlapped Segmentation: EEG signal of duration D is divided into segments of duration d which overlap with a percentage α. In this type of segmentation, each segment starts before the previous one ends where every two successive segments share an overlapping number of data points $o = \alpha * d * sampling frequency$. The total number of segments by considering overlapping $N = \left[\frac{D-d}{d*(1-\alpha)}\right] + 1$. We note that the disjoint segmentation is an overlapped segmentation when $\alpha = 0$.

We proceed to different segmentation assay for a range of segment's duration $S_2 = \{d_1, d_2, ..., d_m\}$ and different overlap's percentages $S_3 = \{\alpha_1, \alpha_2, ..., \alpha_r\}$. Every segmentation $\langle d_i, \alpha_j \rangle$ is carried on each EEG signal of set EEG_{S_1} in order to analyze this segmentation through different level of identity information quality. Formally, for a given segmentation $\langle d_i, \alpha_j \rangle$ as an input, we calculate a set of performance $P = \{P_1, P_2, ..., P_n\}$ returned from EEG-based biometric system using each EEG signal of set $EEG_{S_1} = \{EEG_{f_1}, EEG_{f_2}, ..., EEG_{f_n}\}$.

We hypothesis that the identity information loss comes out in a deterioration in performance of EEG-based biometric system. The proposition of the following metric allows the evaluation of identity information loss:

$$metric\,\langle d_i, \alpha_j \rangle = mean\left(\frac{P_n - P_i}{P_n}\right), 1 \leq i \leq n - 1$$

For each segmentation, the proposed metric presents a global evaluation of identity information loss through the mean deterioration in performance compared to the original performance.

2.3   EEG-Based Biometric System

EEG-based biometric system processes EEG signals related to all subjects to be identified by this system. Formally, we note $EEG = \{EEG_{U_1}, EEG_{U_2}, ..., EEG_{U_E}\}$ where EEG_{U_i} is EEG signal related to subject U_i for $1 \leq i \leq E$ and E is the total number of subjects in the system. Thereby, EEG-based biometric system goes through the following steps:

1. Selection of a set of M channels, the source of information handled by the system, resulting in M EEG signals for every subject broadcasted at the same time. Formally, we note $EEG_{U_i} = \{EEG^1, EEG^2, ..., EEG^M\}_{U_i}$ where $EEG^j_{U_i}$ represents EEG signal of subject U_i resulting from channel j, $1 \leq j \leq M$.

2. Segmentation of all EEG signals as mentioned in the previous section. For a given segmentation $\langle d, \alpha \rangle$, we note:

$$Segment_{U_i} = \begin{pmatrix} Seg^1_1 & \cdots & Seg^M_1 \\ \vdots & \ddots & \vdots \\ Seg^1_N & \cdots & Seg^M_N \end{pmatrix}_{U_i}$$

Where Seg^j_k is the kth segment of the jth channel for subject U_i for $1 \leq k \leq N$ and $1 \leq j \leq M$. We remind that $N = \left[\frac{D-d}{d*(1-\alpha)}\right] + 1$ and $size(Seg^j_k) = d*f$ where f is the sampling frequency of EEG signals.

3. Feature extraction from every segment of EEG signal. In our EEG-based biometric system, we use Autoregressive model for feature extraction described by a linear difference equation in the time domain as follows [9, 15]:

$$x(k) = P + \sum_{i=1}^{P} a(i) x(t-i) + e(t)$$

where P is a constant, P stands for the number of parameters of AR model and $e(t)$ denotes a white noise input. Burg's method is used for estimating the AR model parameters. Hence, we note:

$$Features_{U_i} = \begin{pmatrix} ftr^1_1 & \cdots & ftr^M_1 \\ \vdots & \ddots & \vdots \\ ftr^1_N & \cdots & ftr^M_N \end{pmatrix}_{U_i} \Rightarrow Features_{U_i} = \begin{pmatrix} tmp_1 \\ \vdots \\ tmp_N \end{pmatrix}_{U_i}$$

Where ftr^j_k is the z-score of AR model parameters extracted from segment Seg^j_k and $size\left(ftr^j_k\right) = P$. $Features_{U_i}$ is then transformed by concatenating the kth segments ftr^j_k of all channels for $1 \leq j \leq M$. We note: $tmp_k = ftr^1_k||ftr^2_k||...||ftr^M_k$ for each $1 \leq k \leq N$. Thus, $size(tmp_k) = P*M$.

Therefore, we note the extracted identity information of all subjects of EEG-based biometric system as follows:

$$Features = \begin{pmatrix} tmp_1^1 & \cdots & tmp_1^E \\ \vdots & \ddots & \vdots \\ tmp_N^1 & \cdots & tmp_N^E \end{pmatrix}$$

Where tmp_k^i is the kth extracted identity information representing subject U_i for $1 \leq k \leq N$ and $1 \leq i \leq E$.

4. Classification of all extracted features to return the identity of any subject in EEG-based biometric system. The problem is summed up to a multi-classification problem of E classes using the $E * N$ extracted identity information of $Features$. For this point, we choose K-Nearest Neighbor (KNN) classifier [10,16] which trains and tests features through kfold-crossvalidation. We propose to keep training and test sets disjoint and even not correlated because of overlapped segmentation. Formally, we note:

$$P_x = \begin{pmatrix} tmp_1^1 & \cdots & tmp_1^E \\ \vdots & \ddots & \vdots \\ tmp_T^1 & \cdots & tmp_T^E \end{pmatrix} \quad where \quad T = \left[\frac{N}{kfold} \right] \quad and \quad 1 \leq x \leq kfold$$

For every P_x, tmp_k^i are chosen consecutive from $Features$ to ensure that overlapped parts are grouped in the same part P_x and P_x does not correlate with any other $P_{x'}$ for any $x \neq x'$. Finally, the cross-validation is carried on $D = \{P_x\}_{x=1}^{x=kfold}$. The performance of EEG-based biometric system is the mean of $Performance_x$ for $1 \leq x \leq kfold$ where $Performance_x$ is the performance obtained when considering P_x as a test set.

3   Experiments

We empirically simulate and evaluate the identity information loss through different signal segmentation scenarios in order to find some criteria linked to data segmentation allowing to reduce as far as possible the simulated loss in identity information. In the following, we first describe the datasets used. Then, we present the test scenarios of the methodology presented in Sect. 2. Finally, we analyze and discuss our results.

3.1   Datasets

EEG data from Motor Movement/Imaginary datasets[1] [6] were used in our experiments. Data were collected from 109 healthy volunteers using the BCI2000 System in two different baseline conditions (1-min Eyes Opened resting state; 1-min Eyes Closed resting state). Signals were recorded from 64 channels with a sampling rate of 160 Hz.

---

[1] http://physionet.org/pn4/eegmmidb.

3.2   Test Scenarios of the Methodology

Table 1 summarizes all the test scenarios mentioned in Sect. 2. We test for both
EEG signals in resting state with opened and closed eyes. We select 104 subjects
from the 109 volunteers that have the same signal time length for both EO and
EC.

Table 1. Different test scenarios of the methodology

General	- Original sampling $= 160\,\mathrm{Hz}$
	- Signal duration $\mathrm{D} = 60\,\mathrm{s}$
	- Total number of subjects $\mathrm{E} = 104$
Channel selection	- Selected channels $= \{$FP1, FP2, F3, F4, F7, F8, FZ, CZ, PZ, T3, T4, C3, C4, P3, P4, T5, T6, O1, O2$\}$ [9]. Thus, the total number of channels $\mathrm{M} = 19$
Preprocessing	- The set of sampling frequencies $S_1 = \{80; 100; 120; 140; 160\,\mathrm{Hz}\}$
	- The set of segment's duration $S_2 = \{1; 2; 3; 4; 5; 6; 7; 8; 9; 10\,\mathrm{s}\}$
	- The set of overlap's percentages $S_3 = \{0; 0.25; 0.5; 0.75\}$
Feature extraction	AR model parameters with order $\mathrm{P} = 5$
Classification	- KNN classifier with $\mathrm{K} = 1$ using Euclidean distance [7,16]
	- Kfold-crossvalidation with Kfold $= 3$

3.3   Results and Discussion

We evaluate the loss in identity information in case of disjoint segmentation and
overlapped segmentation for both Eyes-Opened and Eyes-Closed data.

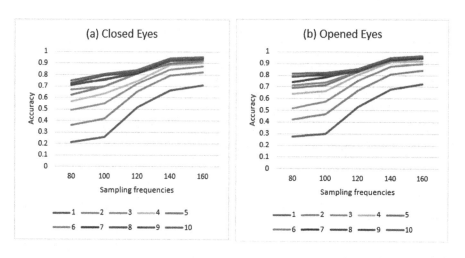

Fig. 2. Accuracy of EEG-based biometric system per sampling frequency for different
disjoint segmentation.

Evaluation of Identity Information Loss Through Disjoint Segmentation. We first simulate the loss in identity information using the different sampling frequencies in S_1. Then, we proceed to different disjoint segmentation scenarios according to segment's duration in S_2. For disjoint segmentation, the percentage of overlap $\alpha = 0$. Results in Fig. 2 show that for both closed and opened eyes cases, the performance of EEG-based biometric system decreases when the quality of identity information is deteriorated (by decreasing the sampling frequency). This emphasizes our hypothesis that the loss in identity information comes out in a deterioration in performance of EEG-based biometric system. Besides, we observe that the deterioration in performance varies strongly from a segmentation to another. Thus, for every segmentation, we calculate the global evaluation of identity information loss mentioned in Sect. 2. Results in Fig. 3(a) show that short segments are more affected by the simulated loss in identity information than longer segments. This can be justified by the fact that longer segments might carry more identity information than shorter ones. However, when segments are longer, the total number of segments strongly decreases as illustrated in Fig. 3(b). Indeed, EEG signal has a limited and fixed duration, longer segments therefore implies less segments. This diminution in number of segments will impact on quality learning of classifier. One solution that may increase the number of segments is the overlapped segmentation. Moreover, Fig. 3(a) shows that segments with duration $d < 4$ are strongly damaged by the simulated loss of identity information whereas segments with duration $d > 7$ are stable against this loss. Thus, we choose segments of duration $S_2' = \{4, 5, 6, 7\}$ as a new area of interest for further segmentation.

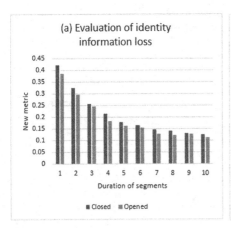 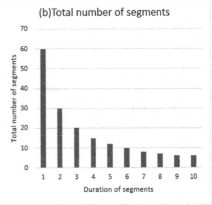

Fig. 3. Disjoint segmentation (a) Evaluation of identity information loss (b) Total number of segments.

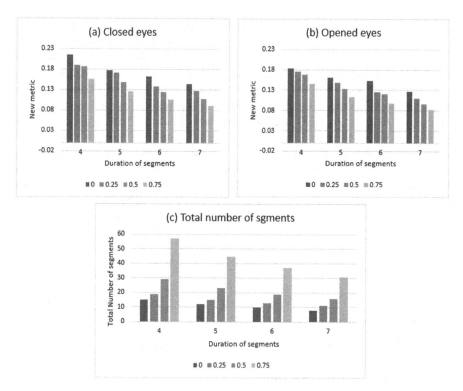

Fig. 4. Overlapped segmentation (a) Evaluation of identity information loss for EC (b) Evaluation of identity information loss for EO (c) Total number of segments.

Evaluation of Identity Information Loss Through Overlapped Segmentation. We proceed to different overlapped segmentation scenarios according to segment's duration in $S_2' = \{4, 5, 6, 7\}$ and overlap's percentages in S_3. Results in Fig. 4(a) and (b) show that overlapped segmentation could reduce the impact of the simulated loss in identity information favoring larger percentages of overlap. Besides, Fig. 4(c) asserts that overlapped segmentation produces more segments compared to disjoint segmentation.

4   Conclusion

In this work, we mainly simulate and evaluate the loss in identity information for EEG-based biometric system through different signal segmentation scenarios in order to find some criteria linked to data segmentation allowing to reduce as far as possible the simulated loss. Our results show that loss in identity information strongly impacts the performance of EEG-based biometric system. Regarding segmentation scenarios, results show that longer segments stands better to the simulated loss in identity information. However, EEG signal has a limited and fixed duration, longer segments therefore implies less segments. Overlapped segmentation addresses this problem by producing more overlapped segments. Besides, overlapped segmentation could reduce the impact of the simulated loss in identity information favoring larger percentages of overlap. In this paper, the loss in identity information is addressed through simulation in order to first emphasize its impact on EEG-based biometric system and then find segmentation scenarios reducing this impact. However, the possibility of loss in identity information is a fact in EEG-based biometric system due to the uncertainty in locating the whole identity information. The evaluation of this loss through different data processing scenarios allows to define criteria of how to locate the identity information as complete as possible in EEG-based biometric system. This evaluation may also identify the real sources of losing identity information in EEG-based biometric system. In our future work, we will address the problem of identity information loss through channel selection and pattern recognition techniques.

References

1. Abo-Zahhad, M., Ahmed, S.M., Abbas, S.N.: State-of-the-art methods and future perspectives for personal recognition based on electroencephalogram signals. IET Biom. **4**(3), 179–190 (2015)
2. Boubakeur, M.R., Wang, G., Zhang, C., Liu, K.: EEG-based person recognition analysis and criticism. In: 2017 IEEE International Conference on Big Knowledge (ICBK), pp. 155–160. IEEE (2017)
3. Campisi, P., La Rocca, D.: Brain waves for automatic biometric-based user recognition. IEEE Trans. Inf. Forensics Secur. **9**(5), 782–800 (2014)
4. Chan, H.L., Kuo, P.C., Cheng, C.Y., Chen, Y.S.: Challenges and future perspectives on electroencephalogram-based biometrics in person recognition. Front. Neuroinform. **12**, 66 (2018)
5. Dmitry, O., et al.: Evolution and evaluation of biometric systems. In: Proceedings of the Second IEEE International Conference on Computational Intelligence for Security and Defense Applications (CISDA 2009), pp. 318–325 (2009)
6. Goldberger, A.L., et al.: PhysioBank, PhysioToolkit, and PhysioNet: components of a new research resource for complex physiologic signals. Circulation **101**(23), e215–e220 (2000)
7. Kaewwit, C., Lursinsap, C., Sophatsathit, P., et al.: High accuracy EEG biometrics identification using ICA and AR model. J. ICT **16**(2), 354–373 (2017)
8. La Rocca, D., Campisi, P., Scarano, G.: EEG biometrics for individual recognition in resting state with closed eyes. In: 2012 BIOSIG-Proceedings of the International Conference of Biometrics Special Interest Group (BIOSIG), pp. 1–12. IEEE (2012)
9. Maiorana, E., La Rocca, D., Campisi, P.: On the permanence of EEG signals for biometric recognition. IEEE Trans. Inf. Forensics Secur. **11**(1), 163–175 (2015)
10. Mao, C., Hu, B., Wang, M., Moore, P.: EEG-based biometric identification using local probability centers. In: 2015 International Joint Conference on Neural Networks (IJCNN), pp. 1–8. IEEE (2015)
11. Mohammadi, G., Shoushtari, P., Molaee Ardekani, B., Shamsollahi, M.B.: Person identification by using ar model for EEG signals. Proceeding World Acad. Sci. Eng. Technol. **11**, 281–285 (2006)

12. Paranjape, R., Mahovsky, J., Benedicenti, L., Koles, Z.: The electroencephalogram as a biometric. In: Canadian Conference on Electrical and Computer Engineering 2001. Conference Proceedings (Cat. No. 01TH8555), vol. 2, pp. 1363–1366. IEEE (2001)
13. Poh, N., Chan, C., Kittler, J., Fierrez, J., Galbally, J.: D3.3: description of metrics for the evaluation of biometric performance. Evaluation 1 (2011)
14. Poulos, M., Rangoussi, M., Chrissikopoulos, V., Evangelou, A.: Parametric person identification from the EEG using computational geometry. In: ICECS 1999, Proceedings of ICECS 1999 6th IEEE International Conference on Electronics, Circuits and Systems (Cat. No. 99EX357), vol. 2, pp. 1005–1008. IEEE (1999)
15. Rodrigues, D., Silva, G.F., Papa, J.P., Marana, A.N., Yang, X.S.: EEG-based person identification through binary flower pollination algorithm. Expert Syst. Appl. **62**, 81–90 (2016)
16. Singh, B., Mishra, S., Tiwary, U.S.: EEG based biometric identification with reduced number of channels. In: 2015 17th International Conference on Advanced Communication Technology (ICACT), pp. 687–691. IEEE (2015)

Weighted-LDA-TVM: Using a Weighted Topic Vector Model for Measuring Short Text Similarity

Xiaobo He[1], Ning Zhong[1,3,4,5(✉)], and Jianhui Chen[1,2]

[1] Faculty of Information Technology, Beijing University of Technology,
Beijing 100124, China
liuruohao@emails.bjut.edu.cn
[2] Beijing Advanced Innovation Center for Future Internet Technology,
Beijing University of Technology, Beijing 100124, China
[3] Beijing International Collaboration Base on Brain Informatics
and Wisdom Services, Beijing 100124, China
[4] Beijing Key Laboratory of MRI and Brain Informatics, Beijing 100124, China
[5] Department of Life Science and Informatics, Maebashi Institute of
Technology, Maebashi, Gunma 371-0816, Japan
zhong@maebashi-it.ac.jp

Abstract. Topic modeling is the core task of the similarity measurement of short texts and is widely used in the fields of information retrieval and sentiment analysis. Though latent dirichlet allocation provides an approach to model texts by mining the underlying semantic themes of texts. It often leads to a low accuracy of text similarity calculation because of the feature sparseness and poor topic focus of short texts. This paper proposes a similarity measurement method of short texts based on a new topic model, namely Weighted-LDA-TVM. Latent dirichlet allocation is adopted to capture the latent topics of short texts. The topic weights are learned by using particle swarm optimization. Finally, a text vector can be constructed based on the word embeddings of weighted topics for measuring the similarity of short texts. A group of text similarity measurement experiments were performed on biomedical literature abstracts about antidepressant drugs. The experimental results prove that the proposed model has the better distinguish ability and semantic representation ability for the similarity measurement of short texts.

Keywords: Similarity measurement of short texts · Topic model · LDA · PSO · Word embedding

1 Introduction

There are many text data on the Internet, such as web pages, XML documents, and e-mails. It has become an urgent task in the field of information technology to effectively manage and analyze these various texts. The similarity measurement of short texts is a core issue of text analysis and mainly calculates semantic distance between texts. It is widely used in the fields of repeated detection of texts [1], information retrieval [2], and sentiment analysis [3].

© Springer Nature Switzerland AG 2019
P. Liang et al. (Eds.): BI 2019, LNAI 11976, pp. 212–219, 2019.
https://doi.org/10.1007/978-3-030-37078-7_21

Topic modeling is the core task of the similarity measurement of short texts. At present, Latent Dirichlet Allocation (LDA) has been one of the most widely used methods for topic modeling. It provides an approach to model texts by mining the underlying semantic topics of texts. However, there are still some limitations. Based on LDA, some textual features, such as word order and grammar, are neglected. The text is simplified into a combination of feature words that are not related to each other. Therefore, this paper proposes a similarity measurement method of short texts based on a new topic model which integrates the weighted LDA and word embedding.

The rest of this paper is organized as follows. Section 2 introduces the related researches. Section 3 gives the details of the proposed method. Experiments are presented in Sect. 4 for validating the superiority of the proposed method. Finally, Sect. 5 gives concluding observations.

2 Related Work

At present, the similarity measurement methods of texts can mainly be divided into three types: string-based methods, knowledge base-based methods and corpus-based methods.

The string-based methods use the original texts themselves to represent texts and measure the similarity by string comparison. This kind of methods is simple and easy to implement. However, because characters or words are used as units independent from knowledge, the string-based methods fail to fully consider the meaning of words and their relationships, and impact the accuracy of similarity measurement. For examples, synonyms have different expressions. Only depending on the string-based methods, their same meaning cannot be considered into the similarity measurement.

The knowledge base-based methods calculate the semantic relevance between two texts by using the information obtained in the knowledge base, such as the hierarchical relationships between concepts in the domain dictionary. The shortcoming of this kind of methods is that the accuracy of similarity measurement depends on the completeness of the domain knowledge base. However, a complete and accurate knowledge base is difficult to be obtained.

The corpus-based methods calculate the similarity between texts based on the information obtained from the large corpora. Their basic idea is to identify a or a group of topic words firstly. Secondly, these topic words are transformed into semantic information vectors by using corpora, and then judge the semantic similarity according to the vector similarity [4–7]. Because corpora are easier to be acquired than knowledge bases, the corpus-based methods have become the most widely used and studies methods in the field of text similarity measurement. The LDA is the most popular method for extracting topic words, and word embedding [8] is the mainstream method for word vectorization.

3 A Similarity Measurement Method of Short Texts Based on Weighted-LDA-TVM

In order to improve the accuracy of similarity measurement of short texts, this paper proposes a method based on a novel weighted topic vector model, namely Weighted-LDA-TVM, which combines LDA with particle swarm optimization (PSO) [9]. The framework of method is shown in Fig. 1. It includes three stages: topic selection, topic weight learning, and text distance calculation.

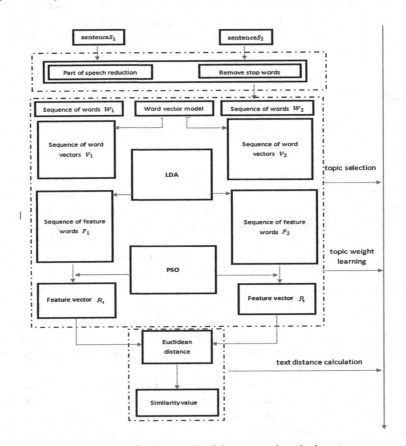

Fig. 1. The framework of the proposed method

3.1 Topic Selection

Topic selection is the first step of proposed method. It includes data preprocessing and topic identification. Data preprocessing mainly includes lemmatization, removing duplicate words and removing stop words. Stanford-Parser can be used to complete these operations. Topic identification is performed by using the following LDA algorithm.

Suppose that the domain corpus set contains D short texts, T topics and T_l unique terms. The ith term t_{li} in any text can be represented as $P(t_{li}) = \sum_{j=1}^{T} p(t_{li}|z_i = j)p(z_i = j)$, where, z_i is the potential variable, indicating that the ith term is taken from the topic; $P(z_i = j)$ gives the probability that topic j belongs to the current text.

Suppose that T topics form D texts represented by T_l unique terms. Let $\varphi_T^{(z)} = p(t_{li}|z_i = j)$ denotes multi-distribution over T_l terms for topic j; let $\partial_{z=j}^{(d)} = p(z = j)$ denotes multi-distribution for text d on T topics, then in text d the probability of the term t_{li} is $p(t_{li}|d) = \sum_{j=1}^{T} \varphi_{t_{li}}^{(z=j)} \cdot \partial_{z=j}^{(d)}$.

The LDA model makes a symmetrical Dirichlet prior probability hypothesis on, $\varphi^{(z)}$ denoted as $\varphi^{(z)} \sim$ Dirichlet (α), and makes a symmetrical Dirichlet prior probability hypothesis on $\theta^{(d)}$, denoted as $\theta^{(d)} \sim$ Dirichlet (β). In this study, Gibbs sampling method is selected to estimate the parameters of $\varphi^{(z)}$ and $\theta^{(d)}$.

According to the formula $p(t_{li}|d) = \sum_{j=1}^{T} \varphi_{t_{li}}^{(z=j)} \cdot \partial_{z=j}^{(d)}$, the lexical probability distribution in the text is obtained. Thereby, the keywords of the corresponding unique terms T_l ($t_{l1}, t_{l2}....t_{lk}$) can be obtained, and the feature word $F_l = (f_{l1}, f_{l2}...f_{lh})$ is selected by calculating the semantic distance between each word and T_l in the short text, where h is the number of feature words, after repeated experiment. In this study, h is set to 10, and the words in F_l are arranged in the descending order according to their semantic distance from the subject words.

Finally, a topic set $T_l(t_{l1}, t_{l2}...t_{lk})$ can be obtained to provide a high-summary expression of the short text.

3.2   Topic Weighting Learning

The feature word set F_l of short texts can be obtained by the trained LDA model, but their importance is different. By using the PSO, this paper introduces the feature weight learning mechanism into topic modeling of short texts.

Suppose that l is a short text and its feature word set $F_l = (f_{l1}, f_{l2}....f_{lh})$. The feature vector corresponding to F_l is ($Vec_l = v_{f_{l1}}, v_{f_{l2}}....v_{f_{lh}}$), in which each $v_{f_{li}}$ is obtained by using the trained Word2Vec model. Then, the text vector of l can be defined as follows.

$$\forall l \in \{1,2\} : V_l = \frac{1}{h}\sum_{g=1}^{h} \beta_g \cdot v_{f_{lg}} \tag{1}$$

where β_g is the weight factor corresponding to $v_{f_{lg}}$.

In order to learn the weight factor β_g, this paper designed a loss function f to measure the relevance between a pair of short texts $<V_i, V_j>$. The greater the f value is, the more relevant the $<V_i, V_j>$ has. This loss function can be defined as follows.

$$F(V) = \begin{cases} d(V_i, V_j), & \text{If V is a semantically related} \\ -d(V_i, V_j), & \text{If V is a semantically unrelated} \end{cases} \tag{2}$$

Where $d(\cdot)$ is a function used to measure the semantic distance between V_i and V_j. It can be defined as follows.

$$d(V_1, V_2) = \sqrt{\sum_{j=1}^{nc} (V_{1j} - V_{2j})^2} \tag{3}$$

Then, the objective function of weight factors can be defined as follows.

$$P(\beta_1, \beta_2 \ldots \beta_h) = \frac{1}{|D|} \sum_{V \in D} F(V) \tag{4}$$

Where D represents the set composed of short text pairs, and $|D|$ represents the total number of short text pairs.

The standard PSO algorithm was adopted to minimize the objective function. The population size is set to 30, the inertia weight is set to 1, the learning factor is set to 2, and the maximum number of iterations is 10000. The change of the weight factor obtained by PSO is shown in Fig. 2. As shown in this figure, with the index of feature words increases, the value of the weight factor decreases steadily. This indicates that the closer the feature word is to the topic word, the larger the weight factor is in the short text.

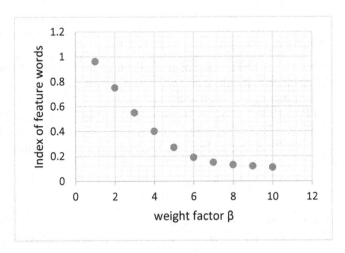

Fig. 2. The change of the weight factor of the characteristic words in the short text

3.3   Text Distance Calculation

By the above steps, the vectors of short texts can be obtained. The semantic distance between two short texts can be calculated by using the Euclidean distance:

$$STS(V_1, V_2) = d(V_1, V_2) \tag{5}$$

4   Experimental Results and Analysis

4.1   The Experimental Dataset

PubMed is a free database of biomedical papers. This study collected PubMed abstracts as the train dataset. The total 2,500 literatures were crawled based on multiple domains of keywords, including "depression", "epilepsy", "cytology", "clinical medicine", and "computer". The abstracts crawled by the keyword "depression" were used to construct the original target corpus set M_1, and the abstracts crawled by the keywords "epilepsy", "cytology", "clinical medicine", and "computer" were used to construct the original contrast corpus M_3. In this paper, we define a natural language sentence as a short text. 10,000 short texts were obtained from M_1 and M_3. Based on these short texts, a semantically related pair set R_1 and a semantically unrelated pair set R_2 were constructed. In R_1, each pair consists of two short texts, all of which were obtained from M_1. In R_2, each pair consists of two short texts in which one was obtained from M_1 and another one was obtained from M_3. In this paper, PSO was trained on R_1 and R_2 to get the optimal weighting factors.

The test dataset was also obtained from PubMed. Similarly, the short texts obtained based on the keyword "depression" were used to construct a target corpus set. The short texts obtained based on the keywords "epilepsy", "cytology", "clinical medicine", and "computer" were used to construct a contrast corpus set. Two corpus sets consist of 5,000 documents. Finally, both the train dataset and the test dataset were used as training corpora for getting the Word2Vec model.

4.2   Experiment and Result Analysis

In this study, two contrast methods were performed to illustrate the effectiveness of proposed method. The first one is the "Mean of the embeddings" method [10], which takes the mean value of all word embeddings in the text as the text vector. The second one is the "Mean+idf weighed" method [10], which chooses the topic words from the text by using the IDF method, and then takes the mean value of weighted word embeddings of topic words as the text vector. Finally, the cosine distance was used to measure the similarity of short texts in both the "Mean of the embeddings" method and the "Mean+idf weighed" method. TF-IDF is a kind of classical method for identifying topics of texts [11] [12]. Experimental results are shown in Fig. 3. In the proposed method, the topic number of LDA is set to 10 and each topic contains 10 words.

As shown in Fig. 3, the target corpus curve and the contrast corpus curve of the "Mean+idf weighed" vector method are close to each other, almost completely

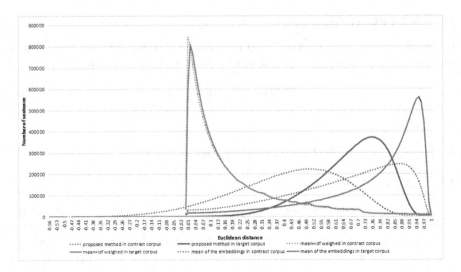

Fig. 3. Experimental results

coincident, and the coincidence degree of the target corpus curve and the contrast corpus curve of the "Mean of the embeddings" method is higher. Compared with the "Mean of the embeddings" method and the "Mean+idf weighed" method, the proposed method can obtain significantly different distance curves on the target corpus set and the contrast corpus set. This means that the proposed method has the stronger semantic representation ability than those two contrast methods.

Furthermore, compared with the curve of "Mean of the embeddings" on the contrast corpus set, the distance curve of the proposal method is mainly located in the regions where similarity values are small. This shows the effectiveness of the weight factor learning based on PSO.

5   Conclusions

Aiming at the shortcomings of LDA in representing the semantics of short texts, this paper proposes a novel weighted-LDA topic vector model for measuring the similarity of short texts. There are two main contributions:

- In order to effectively solve feature sparseness and poor topic focus of short texts, the topic weight is introduced into the LDA method. By using the PSO, the weight factors of topics obtained by the LDA are learning to represent the semantics of short texts more hierarchically and precisely.
- In order to measure the similarity of short texts, the weighted mean value of word embeddings of topics is adopted to construct the text vector. Furthermore, the Euclidean distance is selected to calculate the similarity of short texts based on text vector. The experimental results show that the proposed similarity measurement

method based on the weighted-LDA topic vector can effectively distinguish domain semantics of short texts than the existing contrast methods.

Acknowledgment. The work is supported by Science and Technology Project of Beijing Municipal Commission of Education (No. KM201710005026), National Basic Research Program of China (No. 2014CB744600), Beijing Key Laboratory of MRI and Brain Informatics.

Reference

1. Theobald, M., Siddharth, J., Paepcke, A.: Spotsigs: robust and efficient near duplicate detection in large web collections. In: Proceedings of the 31st Annual International ACM SIGIR Conference on Research and Development in Information Retrieval, pp. 563–570. ACM (2008)
2. Kumar, N.: Approximate string matching algorithm. Int. J. Comput. Sci. Eng. **2**(3), 641–644 (2010)
3. Mohtarami, M., Lan, M., et al.: Sense sentiment similarity: an analysis. In: Twenty-Sixth AAAI Conference on Artificial Intelligence, pp. 1706–1712. AAAI Press (2012)
4. Kenter, T., Rijke, M.D.: Short text similarity with word embeddings. In: ACM International on Conference on Information and Knowledge Management, pp. 1411–1420. ACM (2015)
5. Kusner, M.J., Sun, Y., Kolkin, N.I., et al.: From word embeddings to document distances. In: International Conference on International Conference on Machine Learning, pp. 957–966. JMLR.org (2015)
6. Song, Y., Dan, R.: Unsupervised sparse vector densification for short text similarity. In: Conference of the North American Chapter of the Association for Computational Linguistics: Human Language Technologies, pp. 1275–1280 (2015)
7. Liu, M., Lang, B., Gu, Z., et al.: Measuring similarity of academic articles with semantic profile and joint word embedding. Tsinghua Sci. Technol. **22**(6), 619–632 (2017)
8. Le, Q.V., Mikolov, T.: Distributed representations of sentences and documents, vol. 4, p. II-1188 (2014)
9. Boom, C.D., Canneyt, S.V., Bohez, S., et al.: Learning semantic similarity for very short texts, pp. 1229–1234 (2015)
10. Bafna, P., Pramod, D., Vaidya, A.: Document clustering: TF-IDF approach. In: International Conference on Electrical, Electronics, and Optimization Techniques (2016)
11. Huo, Z., Wu, J., Lu, Y., Li, C.: A topic-based cross-language retrieval model with PLSA and TF-IDF. In: International Conference on Big Data Analysis, pp. 340–344 (2018)

Classification of Mental Arithmetic Cognitive States Based on CNN and FBN

Ruohao Liu[1,2,3,5]([⊠]), Ning Zhong[1,2,3,5,6,7]([⊠]),
Xiaofei Zhang[1,2,3,4,5]([⊠]), Yang Yang[3,5,6,7]([⊠]),
and Jiajin Huang[1,2,3,5,6]([⊠])

[1] Faculty of Information Technology,
Beijing University of Technology, Beijing 100124, China
liuruohao@emails.bjut.edu.cn, zhong@maebashi.ac.jp,
julychang@just.edu.cn, jhuang@bjut.edu.cn
[2] International WIC Institute, Beijing University of Technology, Beijing, China
[3] Beijing Advanced Innovation Center for Future Internet Technology,
Beijing University of Technology, Beijing 100124, China
yang@maebashi-it.org
[4] School of Computer, Jiangsu University of Science and Technology,
Zhenjiang 212003, China
[5] Beijing International Collaboration Base on Brain Informatics and Wisdom
Services, Beijing 100124, China
[6] Beijing Key Laboratory of MRI and Brain Informatics, Beijing 100124, China
[7] Department of Life Science and Informatics,
Maebashi Institute of Technology, Maebashi, Gunma 371-0816, Japan

Abstract. Mental arithmetic is a basic cognitive function of human brain, mental arithmetic is an important cognitive function of brain, which is also considered as the core of human logical thinking. fMRI provides convenience for mental arithmetic cognitive function research because of its non-invasiveness and convenience, More and more experiments are devoted to the clear understanding of mental arithmetic, and the classification of cognitive tasks will contribute to a more comprehensive understanding of the behavior of organisms and the decoding of neural signals. We recruited 21 subjects and took block design of fMRI in the experiment, In this paper, seeds are extracted and characterized by partial correlation connection, build functional brain network (FBN). At present, the deep learning technology represented by CNN is increasingly applied in the analysis and classification of neural image data, CNN was used to classify the brain functional connectivity network, at the same time, with several other machine learning methods classifying mental arithmetic comparison effects and analyzed. The classification results show that in the data-driven classification, CNN-based method have the best classification effect, reaching 98%, which is more obvious than the traditional machine learning method.

Keywords: Mental arithmetic · functional Magnetic Resonance Imaging (fMRI) · Brain functional network · Convolutional neural networks

© Springer Nature Switzerland AG 2019
P. Liang et al. (Eds.): BI 2019, LNAI 11976, pp. 220–230, 2019.
https://doi.org/10.1007/978-3-030-37078-7_22

1 Introduction

Mental arithmetic is one of the advanced cognitive functions of the brain, this mental arithmetic task is to complete the basic addition and subtraction process entirely by means of the function of the brain. Mental arithmetic is considered as a typical cognitive computing and processing task, the basis of artificial intelligence development is to understand the advanced cognitive function of human brain, Number calculation is an important cognitive function of human brain, and it also considered as the core of human logical thinking. The relationship between cognition and computation includes the relationship between the basic unit of cognition and the basic unit of computation [1]. fMRI provides convenience for mental arithmetic cognitive function research because of its non-invasiveness and convenience, More and more experiments are devoted to a clear understanding of the mental arithmetic mechanism. Under the conditions of addition and subtraction, some subsystems work relatively independently, reflecting the separation of brain functional networks [2], and the differences of dominant hemispheres and neural circuits [3]. In recent years, the development of deep learning model in classification and representation learning has greatly promoted the integration of deep learning model and neuroimaging [4]. The multivariate pattern analysis (MVPA) method includes feature selection, extraction, model discrimination, etc. It can be used as a CNN input by using it to analyze multiple variables simultaneously to construct a brain function connection network. There is a growing reliance on machine learning to classify conditions such as depression and Alzheimer's disease, as well as images seen or sounds heard. Mitchell uses machine learning to distinguish between multiple cognitive states such as words, ambiguities and non-ambiguous statements, and pictures [5]. There are also specific brain regions that respond to certain states. and the PPA region and the FFA region respectively consider and process landscape pictures and areas for recognizing faces [6, 7]. Miranda uses SVM and FLD methods to classify faces and locations [8]. Different from the classification of brain diseases with damage in brain regions and regions with specific audiovisual responses, the mental arithmetic of addition and subtraction is a process of brain area coordination, distinguishing addition and subtraction under the condition of calculation, feature selection and classification are difficult.

FBN (functional brain network) refers to the dynamic network connection of neural signals in different time series, which is quantitatively described in the form of correlation coefficient matrix, the purpose is to understand the correlation strength between different brain regions, and it provides an important method for detecting the neural activity patterns in the brain. Generally speaking, research methods of FBN can be divided into two categories. The first category is model-driven, through experiments or prior knowledge, the participating brain regions are determined to clarify the activation region and connection mode of brain regions during calculation, Methods include correlation analysis, partial correlation analysis, etc. The second type is data- driven, without prior knowledge, data-driven functional connectivity patterns of the whole brain were obtained in mental arithmetic experiments, such as principal component analysis and cluster analysis. This experiment by using convolution neural network for the development of machine learning, pattern recognition and so on are applied to the neuroimaging data applications [9–12], complete brain network connectivity and multivariate pattern analysis. Brain network connectivity maps derived from supervised learning can extract features and predict cognitive status. In this paper, the data-driven method is used to classify mental arithmetic and compare its effect. The classification of cognitive tasks will contribute to a more comprehensive understanding of the behavior of organisms and the decoding of neural signals, promote the development of brain-computer interface, and the integration of cognitive science and information science.

2   Materials and Methods

2.1   Subjects

In this experiment, 21 Chinese college students were tested for mental arithmetic, including 12 males and 9 females with an average age of 25.76 years (SD = 3.77). All subjects were right-handed, with a comparable educational background, normal or corrected vision, no neurological disease or a history of mental illness. Before the participants participated in the study, all participants were introduced to the nature of the study and the possible outcomes, and informed consent was obtained from each participant, the study was approved by the Ethics Committee of Xuanwu Hospital of Capital Medical University.

2.2   Experiment Design

fMRI was used in this experiment, involving three modules: addition, subtraction, and number matching. Subjects need to be in accordance with the requirements in the form of photos in order to complete the addition, subtraction calculation and digital matching three tasks, In order to effectively avoid the ceiling effect and the floor effect, the calculation of the addition and subtraction of this test is all a simple calculation of two digits, and does not involve carry and borrow, Addition, subtraction, and number matching are represented by signs of "+", "−", and "#" respectively. The stimulus appears in the form of blocks, In each experiment, the images are presented to the subjects in sequence according to "the first operand", "the second operand", "operation symbol" and "proposed answer", The number and distance of each number varied from 1 to 9, and the frequency of each number was the same in three cases. According to the pictures shown, the subjects pressed the right or left response keys as quickly and accurately as possible. The digital matching module is designed to neutralize the effects of the addition and subtraction modules on the brain to some extent, in which subjects judge whether the proposed answer is one of the two operands.

The block design adopted by the mental arithmetic experiment design scheme has a duration of 6 s for each unit test, and the presentation time of the four operands is 250 ms, 250 ms, 500 ms, and 2000 ms. Each of the two stimuli is separated by a black background of 500 ms. Each experiment interval is 1500 ms. Each block containing 32 experiments, each with a rest block between each block, and the rest block time was 24 s. Figure 1 shows the experimental paradigm.

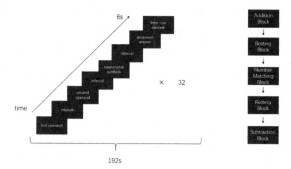

Fig. 1. Experiment paradigm

2.3 Collection and Preprocessing of fMRI Data

The fMRI data was obtained by a German 3.0 Tesla magnetic resonance imaging scanner using a 12-channel phased array head coil. During the experiment, subjects' heads were fixed with foam fillers to reduce the noise generated during the magnetic resonance. Anatomical and functional images were obtained using T1-weighted 3D MPRAGE sequence and EPI sequence, respectively. T1-weighted 3D magnetization method was used to produce fast gradient echo sequence (TR = 1600 ms, TE = 3.28 ms, TI = 800 ms, FOV = 256 mm, flip Angle = 90^0, Voxel size = 1 * 1 * 1 mm^3), and 192 anatomical images with a thickness of 1 mm were obtained. Thirty axial sections with a thickness of 4 mm were obtained.

The subjects' fMRI data were preprocessed by SPM8 software. First, Slice Timing is performed to correct the scan time of the multi-layer image of the whole brain to improve the accuracy of the reconstructed image and improve the temporal correlation. The second is to perform head alignment correction. Thirdly, image registration is carried out to make up for the low spatial resolution of functional image, Meanwhile, the functional part of human brain is matched with the structural part to get the image with high spatial resolution. The fourth step is to perform segment. Fifthly, all volumes were re-sampled to 3 * 3 * 3 mm^3 and standardized. Finally, Smooth is carried out, which can improve the SNR and reduce individual differences.

3 Data-Driven CNN Brain Network Classification

3.1 Construction of FBN

The time series of oxygen level-dependent signals in each ROI region of interest are extracted with MNI coordinates as the center. The signal extracted from all voxels in the region is averaged with a radius of 4.5 mm. Then the time series of each ROI and other correlation coefficients are calculated. Partial correlation is to measure the correlation between two variables while controlling one or more additional variables. Therefore, we choose partial correlation as a multivariate analysis method to find the only certain variance between the two variables. After obtaining the functional connection map, the specific brain region connection coefficient is extracted. The partial correlation matrix is obtained by calculating the correlation matrix and the correlation coefficient R. the formula is as follows:

$$R = \frac{n \sum_{i=1}^{n} x_i y_i - \sum_{i=1}^{n} x_i \sum_{i=1}^{n} y_i}{\sqrt{n \sum_{i=1}^{n} x_i^2 \left(\sum_{i=1}^{n} x_i\right)^2} * \sqrt{n \sum_{i=1}^{n} y_i^2 \left(\sum_{i=1}^{n} y_i\right)^2}} \tag{1}$$

Continue to find the covariance matrix of the correlation matrix and find its inverse matrix, r is the inverse matrix of the covariance matrix of all variables.

$$r = (r_{ij}) = \text{Cov}(R)^{-1} \tag{2}$$

The partial correlation matrix formula is as follows:

$$p = -\frac{r_{ij}}{\sqrt{r_{ii}}\sqrt{r_{jj}}} \tag{3}$$

Many studies divide the brain into different brain regions and form structural templates based on morphological and functional consistency [13–15]. In order to better extract the features between the brain regions for the small world attributes of the brain, we chose the more detailed classification of the Dosenbach-160 template with six sub-networks [16]. The MVPA method selects features with strong discriminating ability from the template and constructs functional connections. The pre-processed file was aggregated into a 4D file through python, which is containing the scan information and time series. After applying the template, each time a 160×160 square partial correlation matrix is created, when the traversal is completed. The time point was 192 s of the total duration of the mental calculation process. All voxels in the template are processed and obtain FBN maps, which are features used in all subsequent convolutional neural networks. Figure 2 shows the FBN when the first subject did addition and subtraction.

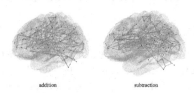

addition   subtraction

Fig. 2. The FBN of subject 1

3.2   Data Enhancement

The number of participants is limited to fewer training samples, so data enhancement is necessary to improve the robustness [17]. The data enhancement methods adopted this time include brightness enhancement, chroma enhancement, contrast enhancement, and sharpness enhancement. In addition, the data set is enriched by noise enhancement.

This experiment uses a method of adding salt and pepper noise and Gaussian noise. Salt and pepper noise, as a kind of noise that often appears in images, appears as a random number of black or white dots appearing in the image. Image simulation adds salt and pepper noise by randomly acquiring pixels and setting them to high and low gray points. For this experiment, 500 black pixels and white pixels are randomly added. The probability density function of Gaussian noise obeys the normal distribution. Contrary to the random noise of salt and pepper noise, the depth of Gaussian noise is randomly generated and distributed in each pixel. At the point, we set the std to 10 on the parameters, the mean value is 0, generate a random Gauss number, and finally output the noisy picture. The Gaussian distribution function is as follows.

$$f(x) = \frac{1}{\sigma\sqrt{2\pi}}\exp\left(-\frac{(x-\mu)^2}{2\sigma^2}\right) \tag{4}$$

While enhancing the data, the robustness of the neural network can also be increased during the training process. In order to reduce the interference term generated by the meaningless connection, the diagonal line (the correlation coefficient between the brain region and itself) is set to zero. In addition, in order to prevent the feature from being reversed, operations such as flipping, mirror rotation, and the like are canceled in the data enhancement. After the combination process, a total of 882 maps were obtained. The overall flow chart is shown in Fig. 3.

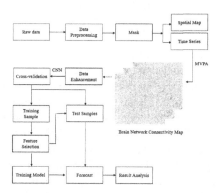

Fig. 3.   Flow chart of experiment

3.3   The Construction of Convolutional Neural Network

The graph data is used as the input of the algorithm, and the original data is converted into the final features through convolution, pooling, activation function mapping and other steps. The CNN network and its parameters are shown in Fig. 4, and the network parameters are listed in Table 1.

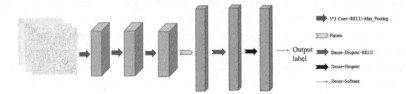

Fig. 4. CNN model

Table 1. CNN parameters

Type	
convolution 1	3*3 strides = [1, 1, 1, 1]
max pool 1	size = [1, 2, 2, 1] | strides = [l, 1, 1, 1]
convolution2	2*2 strides = [1, 1, 1,1]
max pool 2	size = [1, 2, 2, 1] | strides = [1, 2, 2, 1]
convolution3	3*3 strides = [l, 1, 1,1]
max pool3	size = [l, 2, 2, 1] strides = [1, 1, 1, 1]
flatten	
fully connected layer 1	Dense+Dropout+RELU
fully connected layer2	Dense^Dropout
Output	Dense+softmax

In order to unify the input format Settings, the resolution of the graph is 335 * 335, Add and subtract maps are labeled for supervised learning, where the training set is 80% and the validation set is 20%. In order to get the same set of tests for each set of tests when repeated. Therefore, seeds that can generate the same random number are set to facilitate the optimization of the model and improve the accuracy.

The weight is a random value with a standard deviation of 0.05, In order to better fit the data, the performance of the convolutional neural network classification is increased, and the offset of 0.05 is added. Adaptive Moment Estimation is an algorithm that combines the RMSProp optimizer based on Momentum's gradient descent method. It continuously updates the network parameters according to the number of training iterations. The advantage is that each iteration learning rate has a clear range, making the parameter change very stable. In order to make the training converge to a better performance setting, the initial learning rate is $1 * 10^{-4}$, the Adam algorithm has the advantages of high computational efficiency and low memory demand. It is suitable for solving problems involving very high noise or sparse gradients, and it is convenient to adjust parameters, so we choose Adam optimizer. RELU converges faster than Sigmoid and tanh functions due to its linear non-saturation characteristics. The sparse expression ability of the convolutional neural network is improved, which can effectively alleviate the problem of gradient disappearance during training, Activation function select RELU.

In view of the small number of training samples in this cognitive task, Dropout was used to randomly delete 30% neurons in the network and keep the input and output neurons unchanged in order to avoid over-fitting problems due to the large number of parameters in the model. The input parameters are propagated forward through the modified network, and then the network parameters are updated continuously according to the Adam optimizer through the values of the loss function and the network back propagation, so as to prevent excessive dependence on some local features with larger weights. In our architecture, there are three convolution layers, with a subsampling layer based on maximum pooling interpolated in the middle. There will be overlapping areas between adjacent pooling windows, where size is larger than stride. The precision is improved and the over-fitting is not easy to occur. There are two full connection layers in the back. The model was iterated 800 times and the batch size was 32. The input of neurons is mapped to values between 0 and 1 by softmax function. The loss function (i.e. the optimal objective function) selects the cross-entropy function. The cross-entropy loss is calculated by softmax function, and the cross-entropy function is used to measure the distance between the predicted value and the real value. Assume that the probability distribution p is the expected output, the probability distribution q is the actual output, and H (p, q) is the cross entropy to conduct iterative training of the network. The formula is as follows:

$$H(p,q) = -\sum_x (p(x)logq(x) + (1 - p(x))\log(1 - q(x))) \tag{5}$$

4 Results

In this paper, the BP neural network and the traditional machine learning methods KNN and Bayes are selected as comparisons. The classification results are shown in Fig. 5. Compared with fewer samples in model-driven classification, the data-driven classification enlarges the sample size, so the neural network with larger network parameters can be used. Overall, in the data-driven CNN based on its effective feature extraction and good learning ability, the classification accuracy rate is the highest, reaching 98%. The remaining order is BP neural network accuracy rate 74%, Bayes accuracy rate 73%, KNN accuracy rate 62%.

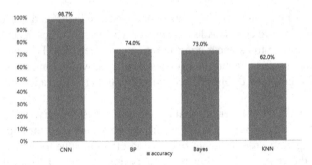

Fig. 5. Classification results of deep learning (CNN, BP) and machine learning (Bayes, KNN)

5 Discussions

The purpose of obtaining the matrix is to convert it into an image, and to fill the color according to the positive and negative feedback, it is easy for the machine to learn and recognize, and the mathematical problem of the rank of the matrix is not considered here. Matrix is the only value when it is used as input. First, it is difficult to expand the sample by adding noise and other methods like the picture. Second, deep learning cannot be carried out to construct the deep learning model. And the purpose of converting the matrix into images: 1. The positive and negative feedback can be easily recognized by the machine by using color areas. (matrix diagrams are generated under the specified toolbox, and the corresponding standard is uniform.) By such representation, a unique matrix image corresponding to the matrix is obtained. 2. The sample can be expanded by a series of methods such as increasing noise and using Generative Adversarial Networks 3. Make use of the advantages of CNN in image processing to identify images with its unique fine-grained feature extraction method, so as to improve the accuracy to a higher level.

BP neural network is a global approximation to nonlinear mapping. The convolution neural network is more suitable for computer vision fields such as image and object recognition. KNN and Bayes are traditional machine learning algorithms. KNN algorithm does not need parameter estimation and training. It classifies by measuring the distance between different eigenvalues. Euclidean distance is selected here. Bayesian algorithm calculates the occurrence frequency of each category in training samples, and estimates the conditional probability of each category by dividing each feature. Finally, classifier is used to classify the samples. The process is simple and fast, but the prediction effect may be poor due to the poor effect of the hypothetical prior model. Generally speaking, CNN and Bp neural networks are better than traditional Bayes and KNN methods. Through data analysis and mental arithmetic classification, using CNN in data-driven classification achieves the highest accuracy. It can get the FBN under mental arithmetic tasks without prior knowledge or experiments. It can also expand samples through data enhancement to facilitate network modification. The disadvantage is that the result interpretation is not intuitive.

6 Conclusions

In this paper, according to the response characteristics of the brain region in the addition and subtraction method, the mental arithmetic tasks are analyzed and classified. Because there are fewer subjects, the optimization of the experiment will further promote the development of mental arithmetic tasks with the increase of sample size in the future. In this paper, three classification methods are compared with CNN method, and the highest classification effect is obtained in CNN. In this data-driven approach, 160 individual elements are taken from fMRI for analysis. Each voxel contains hundreds of thousands or more of nerve cells. Acquiring more precise experimental samples provides space for the development of brain network connection analysis and classification. Algorithms and the choice of different templates have different sizes for classification effects. With the continuous development of network, such as Capsule Network, Residual Networks will better promote the integration of cognitive neuroscience.

References

1. Chen, L.: The core basic science of new generation artificial intelligence: the relation between cognition and computing. J. Chin. Acad. Sci. **33**(10), 108–110 (2018)
2. Zhang, X., Yang, Y., Zhang, M.-H., Zhong, N.: Network analysis of brain functional connectivity in mental arithmetic using task-evoked fMRI. In: Wang, S., et al. (eds.) BI 2018. LNCS (LNAI), vol. 11309, pp. 141–152. Springer, Cham (2018). https://doi.org/10.1007/978-3-030-05587-5_14
3. Yang, Y., et al.: The functional architectures of addition and subtraction: Network discovery using fMRI and DCM. Hum. Brain Mapping **38**(6), 3210 (2017)
4. Plis, S.M., Hjelm, D.R., Salakhutdinov, R., et al.: Deep learning for neuroimaging: a validation study. Front. Neurosci. **8**(8), 229 (2014)
5. Mitchell, T.M., Hutchinson, R., Niculescu, R.S., et al.: Learning to decode cognitive states from brain images. Mach. Learn. **57**(1–2), 145–175 (2004)
6. Epstein, R., Kanwisher, N.: A cortical representation of the local visual environment. Nature **392**(6676), 598–601 (1998)
7. Allison, T., Ginter, H., Mccarthy, G., et al.: Face recognition in human extrastriate cortex. J. Neurophysiol. **71**(2), 821–825 (1994)
8. JanainaMourão-Miranda, J., Bokde, A.L.W., Born, C., et al.: Classifying brain states and determining the discriminating activation patterns: support vector machine on functional MRI data. Neuroimage **28**(4), 980–995 (2005)
9. Simonyan, K., Zisserman, A.: Very deep convolutional networks for large-scale image recognition arXiv:1409.1556 (2014)
10. Szegedy, C., Liu, W., Jia, Y., et al.: Going deeper with convolutions. In: Proceedings of the IEEE Conference on Computer Vision and Pattern Recognition. IEEE (2015)
11. He, K., Zhang, X., Ren, S., et al.: Deep residual learning for image recognition. In: Proceedings of the IEEE Conference on Computer Vision and Pattern Recognition. IEEE (2016)
12. Hu, J., Shen, L., Sun, G.: Squeeze-and-excitation networks arXiv:1709.01507 (2017)

13. Desikan, R.S., Segonne, F., Fischl, B., et al.: An automated labeling system for subdividing the human cerebral cortex on MRI scans into gyral based regions of interest. NeuroImage **31**, 968–980 (2006)

14. Yeo, B.T., Krienen, F.M., Sepulcre, J., et al.: The organization of the human cerebral cortex estimated by intrinsic functional connectivity. J. Neurophysiol. **106**, 1125–1165 (2011)

15. Rolls, E.T., Joliot, M., Tzourio-Mazoyer, N.: Implementation of a new parcellation of the orbitofrontal cortex in the automated anatomical labeling atlas. NeuroImage **122**, 1–5 (2015)

16. Dosenbach, N., et al.: Prediction of individual brain maturity using fMRI. Science **329** (5997), 1358–1361 (2010)

17. Krizhevsky, A., Sutskever, I., Hinton, G.E.: Imagenet classification with deep convolutional neural networks. In: Advances in Neural Information Processing Systems (2012)

Special Session on Computational Social Analysis for Mental Health

Establishment of Risk Prediction Model for Retinopathy in Type 2 Diabetic Patients

Jianzhuo Yan, Xiaoxue Du[(⊠)], Yongchuan Yu, and Hongxia Xu

Faculty of Information Technology, Beijing University of Technology,
Beijing 100124, China
S201761702@emails.bjut.edu.cn

Abstract. Diabetic retinopathy (DR) is one of the complications of diabetes mellitus, which is an important manifestation of diabetic microangiopathy and major cause of vision loss in middle-aged and elderly people worldwide. Establishing a risk prediction model for diabetic retinopathy can discover high-risk groups and early warn diabetic retinopathy, which can effectively reduce the medical cost of diabetes. The experimental data was derived from the electronic medical records of a tertiary hospital of Beijing from 2013 to 2017, including 29 inspection indicators. In this study, we compared the predictive models of type 2 diabetes mellitus complicated with retinopathy, and finally selected the random forest method to construct the risk prediction model. The weights of each index are analyzed by linear regression algorithm, the combination of inspection indicators with the highest accuracy is selected, and the random forest model is optimized to improve the accuracy of the classification prediction model, accuracy increased by 3.7264%. The predictive model provides a basis for early diagnosis of diabetic retina and optimization of the diagnostic process.

Keywords: Type 2 diabetic retinopathy · Risk prediction model · Random forest algorithm · Linear regression analysis

1 Introduction

Diabetic Mellitus (DM) is a metabolic disease characterized by hyperglycemia. In recent years, the incidence of diabetes is increasing, and it has become one of the most serious public health problems in the world of 21st century. The number of adults with diabetes has increased from 382 million in 2013 to 415 million, and is expected to reach 642 million worldwide by 2040 [1]. Similarly, according to China's 2013 edition of the "Guidelines for the Prevention and Treatment of Type 2 Diabetes" data show that 40%–80% of China's adult patients with type 2 diabetes may have eye disease, 8% of patients even blind [2]. If diabetic patients can discover and get standard treatment in time, most of them can get rid of the risk of blindness. Therefore, it is necessary to construct a risk prediction model for diabetes mellitus complicated with retina, which is helpful to predict the individual risk and identify the high risk factors of diabetes complicated with retina as early as possible, so as to improve the health condition and reduce the tremendous economic pressure of patients simultaneously [3].

© Springer Nature Switzerland AG 2019
P. Liang et al. (Eds.): BI 2019, LNAI 11976, pp. 233–243, 2019.
https://doi.org/10.1007/978-3-030-37078-7_23

In recent years, there are many researches on the risk prediction model of diabetic complications. It can be used to predict the possibility of various complications and the occurrence sequence of diabetic patients after several years [4]. The UKPDS Models, a well-known model for predicting diabetes mellitus, simulates the probability and timing of seven common complications and related deaths to find if strict glycemic control reduces the risk of complications in type 2 diabetes mellitus, and is now widely used in diabetes mellitus [5]. Another diabetes model was established by the Basel Institute for Research in Switzerland to simulate the progression of diabetes and its complications – the CORE diabetes model [6]. Besides, many domestic researchers also have studied diabetes complications [7]. Ge used learning vector quantization network (LVQ) to construct the risk prediction model of diabetic complications. He predicted five kinds of diabetic chronic complications with the accuracy of 64.71%– 82.35% [8]. Wang used CHAID method to establish a decision tree prediction model to predict the risk of kidney disease in 369 patients with type 2 diabetes mellitus. Three risk factors were screened out, which were development of disease, gender and systolic blood pressure. The prediction accuracy rate was 74.3% [9]. Therefore, The high-accuracy diabetes risk prediction model can effectively alleviate the shortage of medical resources and uneven distribution of medical resources.

2   Data Source and Method

2.1   Data Source

This study selected diabetes diagnosis data, diabetes glycation test data and biochemical examination data of inpatients in a third-grade hospital in Beijing from 2013 to 2018. The data of 1440 diabetic retinas and 3680 diabetic non-retinal retinas in 2013–2018 were derived from HIS system by SQL. And integrate them to get data sets that can be used for statistical analysis. In addition, some factors are abstracted and transformed from textual case report.

The research model is implemented through the sklearn Library of Python language. The experimental data mainly include 29 items (age, sex, kidney disease, glycosylated hemoglobin, etc.) as shown in Table 1 below.

Table 1. Diabetic retinal related indicators

Basic information	Laboratory indicators		
Age	Hba1c	SCr	TCHO
Gender	ALT	UCr	CK
Hyperlipidemia	AST	ESR	LDH
Hypertension	TP	HDL_C	Ca
Nephropathy	TBIL	Urea N	Na
	DBIL	Mg	K
	ALP	TG	Cl
	S_UA	LDL_C	A/C

At the end of the data, each sample was labeled Outcome, and the Outcome label for diabetic retinopathy was 1. For diabetic patients without diabetic retinopathy, the Outcome label was 0. The number of diabetic patients with diabetic retinopathy labeled 0 was 1400 and that with diabetic retinopathy labeled 1 was 429. For these data, we first analyzed the correlation between 29 indicators and diabetic retina, and compared the results according to the absolute value of the correlation in descending order, and compiled the overall data source (1400) with the data of 1829 subjects. Subjects, 429 diagnosed patients, as shown in Table 2.

Table 2. Indicator correlation table

Item	Correlation coefficient	Mean ± Std. patient (805)	Mean ± Std. normal (796)
DR	1	1 ± 0	0 ± 0
A/C	0.2781	4.833 ± 0.716	4.511 ± 0.388
Hba1c	0.2478	7163.131 ± 1222.016	6580.079 ± 1119.073
gender	0.1919	20.689 ± 16.575	15.554 ± 8.356
Age	0.1894	30.699 ± 4.187	29.146 ± 3.938
Nephropathy	0.1847	9.353 ± 2.214	8.565 ± 1.963
AST	0.1764	6.872 ± 1.896	6.228 ± 1.723
TP	0.1728	267.994 ± 41.306	254.005 ± 39.178
DBIL	0.1717	225.193 ± 55.409	205.731 ± 52.435
SCr	0.1651	4.34 ± 0.355	4.23 ± 0.351
ALT	0.1486	21.291 ± 19.316	16.445 ± 13.025
TCHO	0.146	0.425 ± 0.13	0.389 ± 0.109
LDH	0.1448	38.533 ± 2.503	37.838 ± 2.639
Na	0.1286	268.92 ± 55.52	254.704 ± 53.962
K	0.1224	130.879 ± 9.693	128.618 ± 10.062
ESR	0.1151	0.247 ± 0.047	0.237 ± 0.046
Hyperlipidemia	−0.0458	24.299 ± 1.995	24.495 ± 1.842
Hypertension	−0.0468	6.387 ± 0.614	6.447 ± 0.602
ALT	−0.0499	0.253 ± 0.121	0.265 ± 0.13
TBIL	−0.0536	11.56 ± 3.788	12.568 ± 12.148
ALP	−0.0566	9.273 ± 0.881	9.382 ± 0.965
Urea N	−0.0626	30.241 ± 1.93	30.475 ± 1.872
UCr	−0.0631	0.979 ± 0.059	0.989 ± 0.093
TG	−0.0639	0.979 ± 0.057	0.989 ± 0.09
S_UA	−0.064	11.255 ± 0.674	11.374 ± 1.073
CK	−0.0658	89.037 ± 4.87	89.678 ± 4.917
Ca	−0.075	3.403 ± 1.278	3.64 ± 1.548
Cl	−0.084	6.889 ± 1.202	7.114 ± 1.262
Mg	−0.085	21.27 ± 4.977	22.105 ± 5.403
HDL_C	−0.0859	27.774 ± 7.798	29.111 ± 7.702
LDL_C	−0.183	27.485 ± 3.538	28.731 ± 3.271

2.2 Diabetic Retina Prediction Model

The nearest neighbor (K-Nearest Neighbor, KNN) algorithm is a well-known pattern recognition statistical method and occupies a considerable position in machine learning classification algorithms. KNN is classified by measuring the distance between different eigenvalues. The idea is that if the majority of the k most similar samples in a feature space belong to a certain category, then the sample also belongs to this category, where K is usually an integer no larger than 20. In the KNN algorithm, the selected neighbors are all objects that have been correctly classified. KNN determines the category to which the sample to be classified belongs based on the category of the nearest neighbor or samples. Logistic regression algorithm is a generalized linear regression analysis model, which is often used in data mining, automatic disease diagnosis, economic forecasting, etc. [10]. It is the most commonly used algorithm in statistics. It has the advantages of fast speed, easy to understand, easy to update the model to absorb new data, etc. One result of the two-category response variable Y is referred to as "diabetic retinopathy", indicated by 1; the other result is referred to as "diabetic non-retinal lesion", indicated by 0. If 29 independent variables (influencing factors) $X_1, \cdots, X_{29}$ have an effect on Y at $P = P(Y = 1 | X_1, \cdots, X_{29})$, P is called Logistic regression model P:

$$P = \frac{\exp(\beta_0 + \beta_1 X_1 + \cdots + \beta_{29} X_{29})}{1 + \exp(\beta_0 + \beta_1 X_1 + \cdots + \beta_{29} X_{29})}$$

Decision tree is a commonly used tree structure prediction model for case classification, and it is a supervised learning method. It uses a certain number of given samples, and each sample has a set of attributes and a pre-determined category. Then it learns a classifier through the decision tree algorithm [11]. The classifier can give the correct classification of the new data. The decision tree is composed of nodes and directed edges. There are two types of nodes: internal nodes and leaf nodes. Among them, the internal node represents the test condition of a feature or attribute, and the leaf node represents a classification. Once we construct a decision tree model, it will be very easy to classify it based on it [12]. Specifically, starting from the root node, testing the instance according to a certain feature of the instance, assigning the instance to its child nodes according to the test structure, and recursively executing the new test condition until a leaf node is reached along the branch that may reach the leaf node or another internal node. When we reach the leaf nodes, we get the final classification results [13].

Random forest is an algorithm that integrates multiple trees by ensemble learning. Its basic unit is decision tree. The concrete steps of the stochastic forest model constructed in this study are as follows:

(1) The number of samples in the experimental training set is 1371, and then these 1371 samples are obtained by repeated sampling with reset. Such sampling results will be used as the training set for our decision tree generation [14].

(2) There are 29 input variables, each node will randomly select m (m < M) specific variables, and then use these m variables to determine the best splitting point. In the process of decision tree generation, the value of M remains unchanged.

(3) Every decision tree is most likely to grow without pruning.

(4) The model predicts the new data by adding all the decision trees [15].

3 Results and Discussion

Precision, recall, F1 and accuracy were used as evaluation indexes of the prediction model. Let's suppose a two classification problem with two positive and negative samples [16]. Then there are four combinations of predicted results and real tags: TP, FP, FN, TN, as shown in the Table 3 below. These four indicators are: the actual positive sample forecast is positive sample, the actual negative sample forecast is positive sample, the actual positive sample forecast is negative sample, the actual negative sample forecast is negative sample [17].

Table 3. Four combinations

	True	
Predict	True Positive	False Positive
	False Negative	True Negative

According to the KNN model classification prediction accuracy Fig. 1, this paper takes 7 as the best value of K value, K value range (1, 10). The abscissa is the value of N, the ordinate is the accuracy, the blue is the accuracy of the training set, and the yellow is the accuracy of the test set. This study ensures that the accuracy of the training set is higher, and the accuracy of the test set is higher.

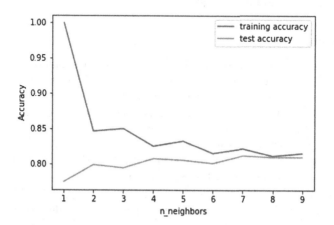

Fig. 1. KNN model classification prediction accuracy graph

In this paper, KNN algorithm is used to predict the diabetic retina model. The idea is to input test data when the training set data and labels are known, compare the characteristics of test data with the corresponding features of training set, and find the first seven data which are most similar to the training set. The classification is the most frequent occurrence in the 7 data. For the two classification problem of this study, the Logistic regression algorithm contains 29 independent variables X. The results of the four models are shown below (Table 4):

Table 4. Model results

Model	Accuracy	Recall	F1	Precision
KNN	80.991 ± 2.833	81.598 ± 2.210	78.264 ± 2.128	79.8725
Logistic	80.707 ± 2.182	81.018 ± 2.861	78.837 ± 2.369	81.7295
DT	75.988 ± 4.155	79.504 ± 2.202	76.549 ± 2.234	76.9971
RF	82.875 ± 1.726	82.644 ± 2.336	81.212 ± 1.705	85.1032

Comprehensive evaluation of the results of a variety of indicators found that the random forest model can find hidden classification rules in the data, the comprehensive effect of the prediction model is better. The recall rate and sensitivity of decision tree model reflect that the model can detect positive results strongly; the probability of classifying negative results as positive results is better by random forest algorithm; F value and accuracy are obviously higher than other algorithms, reflecting the strong consistency between the predicted results of random forest algorithm and the actual situation [18]. The results show that the four algorithms have significant differences, and the random forest algorithm has better classification and prediction effect than other algorithms. Therefore, this study first used a random forest to establish a preliminary predictive model of diabetic retinopathy.

3.1   Index Importance Score

Random forest can construct disease prediction model. In addition, random forest algorithm also has the function of evaluating the importance of variables. The higher the importance score of variables, the more capable the variable is to classify the outcome variables. By calling the feature_importances module of the random forest model, we can get the importance of each input index on the output index [19].

We assume that there are bootstrap samples $b = 1, 2, \ldots B$, B denotes the number of training samples, and the variable importance measure D of feature X J based on classification accuracy is calculated in the following steps [20].

(1) Set up $b = 1$, create decision tree Tb on training samples, and label the data outside the bag as L_b^{oob}. (2) Tb is used to classify L_b^{oob} data on the out of pocket data, and the number of correct classification is recorded as R_b^{oob}. (3) The eigenvalues in L_b^{oob} are perturbed and the perturbed data set is recorded as L_{bj}^{oob}. The L_{bj}^{oob} data is classified by Tb. The number of correct classifications is counted as R_{bj}^{oob}.

(4) Repeat steps (1)–(3). (5) The characteristic X_j variable importance metric D_j is calculated by the following formula.

$$D_j = \frac{1}{B} \sum_{i=1}^{B} R_b^{oob} - R_{bj}^{oob}$$

Through the above calculation method, we can get the importance weight of each feature as shown in Fig. 2 below:

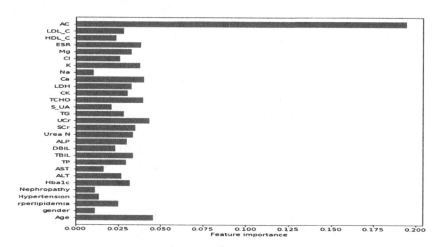

Fig. 2. Each factor importance score

The abscissa is the weight of the feature importance weight, and the ordinate is the variable in the model. The larger the weight ratio, the greater the influence of the corresponding variable on the classification of the model. It can be seen from the figure that the A/C score is significantly higher than other indicators, which plays a significant role in the classification of diabetic retinopathy and diabetic non-retinopathy.

3.2 Optimization of Random Forest Model

we mention that the parameters affecting the accuracy of the random forest model are the depth of the tree and the tree. Next we tuned these two parameters. The optimization parameter method we used is the (GridSearchCV) grid Optimization method. The specific method is as follows:

1. Grid search: First set a set of candidate values for the parameters you want to adjust, then the grid search will exhaust the various parameter combinations, and find the best set according to the set scoring mechanis.

2. Cross-validation: The data classification adopts the method of cross-validation. The general idea of K-time cross-checking is to roughly divide the data into K sub-samples, take one sample each time as verification data, and take the remaining K − 1 samples as training data. (Here we use 5 times cross-validation) [21, 22].

3.3   Logistic Single Factor Regression Analysis of Risk Factors

In this study, A/C single factor score was significantly higher than other factors, we used linear regression analysis on the other 28 indicators for linear regression single factor analysis, screening out other risk factors affecting A/C. Where coef indicates the degree of correlation, less than zero is a negative correlation, and greater than zero is a positive correlation. The coef values of the 28 indicators are as follows (Table 5):

Table 5. Related analysis

Number	Coef	Columns	Number	Coef	Columns
4	0.29169577	Nephropathy	27	−0.00761472	LDL_C
10	0.13375797	DBIL	15	−0.01606504	TG
17	0.09280147	TCHO	26	−0.02412406	HDL_C
0	0.06136515	Age	11	−0.02506173	ALP
13	0.04308944	SCr	24	−0.02576717	Mg
8	0.04039838	TP	16	−0.03369948	S_UA
25	0.03970274	ESR	18	−0.03754701	CK
19	0.03759716	LCH	14	−0.03804691	UCr
21	0.03384397	Na	20	−0.04712560	Ca
5	0.02965216	Hba1c	6	−0.05972014	ALT
7	0.02691554	AST	12	−0.06480643	Urea N
22	0.02643127	K	2	−0.08354605	Hyperlipidemia
1	0.01190249	gender	3	−0.11723897	Hypertension
23	−0.00032104	Cl	9	−0.13872081	TBIL

In this study, a positive correlation with A/C and A/C combination were used to establish a predictive model of diabetic retina. Specific steps are as follows:

1. Select all the indicators that are positively related to the A/C indicator, and sort them according to the coef value from large to small;
2. Combine the indicators with the largest A/C and coef values as risk factors and construct the diabetic retina prediction model using the optimized random Mori algorithm mentioned above;
3. Combine the first two indicators in the A/C and coef value ranking tables as risk factors to construct diabetic retinopathy and so on, and find the best model;

The experimental results are shown in the following Table 6:

Table 6. Indicator combination result

Index	Precision
A/C, Nephropathy	84.1056
A/C, Nephropathy, DBIL	84.1659
A/C, Nephropathy, DBIL, TCHO	84.8796
A/C, Nephropathy, DBIL, TCHO	85.1878
A/C, Nephropathy, DBIL, TCHO, Age	86.3878
A/C, Nephropathy, DBIL, TCHO, Age, SCr	88.8296
A/C, Nephropathy, DBIL, TCHO, Age, SCr, TP	87.8198
A/C, Nephropathy, DBIL, TCHO, Age, SCr, TP, ESR	86.8878
A/C, Nephropathy, DBIL, TCHO, Age, SCr, TP, ESR, LDH	85.1004
A/C, Nephropathy, DBIL, TCHO, Age, SCr, TP, ESR, LDH, Na	83.8103

From the data in the table, the accuracy of the model is 84.1056% when A/C and Nephropathy are selected as risk factors. The precision rate increases with the increase of the selected index, and A/C, Nephropathy, DBIL, TCHO, Age, SCr are selected in the risk factors. When the six indicators are used, the model precision rate is the highest, and then the accuracy of the index model is reduced. Therefore, this study selected A/C, Nephropathy, DBIL, TCHO, Age, and SCr as risk factors to construct a model using the optimized random forest algorithm.

In the diabetic retinopathy model constructed by other algorithms, it was found that the random forest model prediction model was better than other models. In this study, the random forest model was optimized and risk factors were combined to construct a predictive model of diabetic retinopathy, which improved the accuracy of the model. Six indicators of A/C, Nephropathy, DBIL, TCHO, Age and SCr were used as clinical guidance and risk assessment.

4 Conclusions

Based on the review of relevant theoretical research at home and abroad, this study proposes a strategy for predicting the model of retinopathy in diabetic patients, and conducts experimental research on the construction of disease prediction models. In this study, KNN, Logistic, decision tree, and random forest were used to analyze the risk factors of diabetic retinopathy. By contrast, we selected a random forest algorithm with higher comprehensive index to construct the model and optimize the random forest model. Finally, A/C, Nephropathy, DBIL, TCHO, Age, SCr were selected as risk factors to optimize the random forest model to construct the prediction model. Increasing the accuracy rate to 88.8296% provides a reference for the data preprocessing and model construction of medical big data in theory, and provides reliable decision support for the prevention, diagnosis and treatment of diabetic retinopathy in practice.

Acknowledgements. This work is supported by the CERNET Innovation Project (No. NGII20170719) and the Beijing Municipal Education Commission.

Declarations.

Authors' Contributions

Developed the method: JY, XH and YY. Conceived and designed the experiments: XD. Analyzed the data: XD and JY. Wrote the first draft of the manuscript: XD. Contributed to the writing of the manuscript: XD, JY, XH and YY. Agree with the manuscript results and conclusions: JY, YY and XD Jointly developed the structure and arguments for the paper: JY, YY and XD. Made critical revisions and approved final version: JY, XD and YY. All the authors reviewed and approved of the final manuscript.

References

1. Field, R.A.: The current status of hormonal suppression in the treatment of diabetic retinopathy. Trans. – Am. Acad. Ophthalmol. Otolaryngol. **72**(2), 241–245 (1968)
2. Zheng, Z.: Clinical prevention and treatment of diabetic retinopathy: progress, challenges and prospects. Chin. J. Fundus Dis. **28**(3), 209–213 (2012)
3. Klein, B.E.: Overview of epidemiologic studies of diabetic retinopathy. Ophthalmic Epidemiol. **14**(4), 179–183 (2007)
4. Nelson, R.G., Newman, J.M., Knowler, W.C.: Incidence of end-stage renal disease in type 2 (non-insulin-dependent) diabetes mellitus in Pima Indians. Diabetologia **31**(10), 730–736 (1988)
5. Levy, J.C., Cull, C.A., Stratton, I.M.: The UKPDS study on glycemic control and arterial hypertension in type II diabetes: objectives, structure and preliminary results. J. Annu. Diabetol. Hotel Dieu (1993)
6. Mcewan, P., Foos, V., Palmer, J.L.: Validation of the IMS CORE diabetes model. Value Health J. Int. Soc. Pharmacoeconomics Outcomes Res. **17**(6), 714–724 (2014)
7. Brändle, M., Herman, W.H.: The CORE diabetes mode. Curr. Med. Res. Opin. **20**(sup1), S1–S3 (2004)
8. Ge, L.I., Jin, L.Z.: Establishing a model for predicting diabetes complications based on the LVQ neural work. Chin. J. Nat. Med. (2006)
9. Wang, Z., Song, Z., Bai, J.: Decision tree analysis of nephropathy risk in patients with type 2 diabetes mellitus. Chin. J. Integr. Tradit. West. Med. Nephrol. **14**(3), 238–239 (2013)
10. Geng, L., Li, X.: Research on KNN algorithm for big data classification. J. Comput. Appl. **31**(5), 1342–1344 (2014)
11. Safavian, S.R., Landgrebe, D.: A survey of decision tree classifier methodology. IEEE Trans. Syst. Man Cybernet. **21**(3), 660–674 (2002)
12. Freund, Y., Mason, L.: The alternating decision tree learning algorithm. In: Proceeding of the, International Conference on Machine Learning, pp. 124–133. Morgan Kaufmann (1999)
13. Mingers, J.: An empirical comparison of pruning methods for decision tree induction. Mach. Learn. **4**(2), 227–243 (1989)
14. Breiman, L.: Random forests. Mach. Learn. **45**(1), 5–32 (2001)
15. Cutler, A., Cutler, D.R., Stevens, J.R.: Random forests. Mach. Learn. **45**(1), 157–176 (2004)
16. Fawcett, T.: An introduction to ROC analysis. Pattern Recogn. Lett. **27**(8), 861–874 (2005)
17. Metz, C.E.: Basic principles of ROC analysis. Semin. Nucl. Med. **8**(4), 283–298 (1978)

18. Genuer, R.: VSURF: variable selection using random forests. Pattern Recogn. Lett. **31**(14), 2225–2236 (2016)
19. Biau, G.: Analysis of a random forests model. J. Mach. Learn. Res. **13**(2), 1063–1095 (2010)
20. Archer, K.J., Kimes, R.V.: Empirical characterization of random forest variable importance measures. Elsevier Science Publishers B. V. (2008)
21. Lindner, C., Bromiley, P.A., Ionita, M.C.: Robust and accurate shape model matching using random forest regression-voting. IEEE Trans. Pattern Anal. Mach. Intell. **37**(9), 1862–1874 (2015)
22. Palmer, D.S., O'Boyle, N.M., Glen, R.C.: Random forest models to predict aqueous solubility. J. Chem. Inf. Model. **47**(1), 150 (2007)

Anesthesia Assessment Based on ICA Permutation Entropy Analysis of Two-Channel EEG Signals

Tianning Li[1](✉), Prashanth Sivakumar[2], and Xiaohui Tao[2]

[1] Faculty of Health, Engineering and Sciences,
University of Southern Queensland, Toowoomba, QLD 4350, Australia
Tianning.Li@usq.edu.au
[2] University of Southern Queensland, Toowoomba, QLD 4350, Australia

Abstract. Inaccurate assessment may lead to inaccurate levels of dosage given to the patients that may lead to intraoperative awareness that is caused by under dosage during surgery or prolonged recovery in patients that is caused by over dosage after the surgery is done. Previous research and evidence show that assessing anesthetic levels with the help of electroencephalography (EEG) signals gives an overall better aspect of the patient's anesthetic state. This paper presents a new method to assess the depth of anesthesia (DoA) using Independent Component Analysis (ICA) and permutation entropy analysis. ICA is performed on two-channel EEG to reduce the noise then Wavelet and permutation entropy are applied on these channels to extract the features. A linear regression model was used to build the new DoA index using the selected features. The new index designed by proposed methods performs well under low signal quality and it was overall consistent in most of the cases where Bispectral index (BIS) may fail to provide any valid value.

Keywords: Depth of anesthesia · Electroencephalograph · Independent component analysis · Permutation entropy

1 Introduction

Anesthesia is a drug that is used on patients during a medical operation such as surgery to reduce the pain and discomfort that patients may face during surgery. The anesthesia depth (DoA) is reflected in the change of partial pressure of anesthetics in the brain [1]. It is important to obtain accurate DoA assessment results since inaccurate assessment may lead to inaccurate levels of dosage given to the patients that may lead to intra-operative awareness that is caused by under dosage during surgery or prolonged recovery in patients that is caused by over dosage after the surgery is done [2, 3]. Previous research and evidence show that assessing anesthetic levels with the help of electroencephalography (EEG) signals gives an overall better aspect of the patient's anesthetic state. Bispectral index (BIS) is the widely used monitor today in hospitals but it has a couple of issues such as failing to give accurate levels of anesthesia when the signal quality is low and providing inaccurate assessment for certain patients with medical issues [4]. The existing algorithms that are present today that try to solve the

© Springer Nature Switzerland AG 2019
P. Liang et al. (Eds.): BI 2019, LNAI 11976, pp. 244–253, 2019.
https://doi.org/10.1007/978-3-030-37078-7_24

problems from the bispectral index usually suffer from problems when used in real-time such as time delay and inaccurate data filtering techniques.

One of the problems with the EEG data is there are a lot of corrupt signals or a mix up of signals so interpreting the output and proceeding to the next stage of the process is very difficult. So, one solution to this problem can be blind source separation and the official algorithm to solve this is independent component analysis (ICA). One of the papers that were researched [5] combines both the technique Blind source separation (BSS) and ICA and gives a very decent output where the ECG artifacts are removed from the EEG. Ansell and Hossain introduce a technique that combines ICA and Wavelet to preprocess the EEG data before going to feature extraction [6].

For an accurate and reliable depth of anesthesia assessment, intensive research has been conducted, and various algorithms were developed. The latest methods includes Entropy [7], Detrended moving-average (DMA) [8], Isomap-based estimation [9], Empirical-mode decomposition (EMD) [3], and Bayesian [10]. Nguyen-Ky developed a new technique to rectify the problems of BIS by developing a wavelet-based depth of anesthesia (WDoA) with a help of discrete wavelet transform (DWT) and power spectral density function (PSD) [1]. On comparing the readings of the BIS to real-time, it was found that there was a slight time lag from the BIS monitor. Although the outcome of the paper solves a time lag problem, it doesn't address some serious issues that are possessed by BIS such as missing readings where the BIS fails to give any valid output and handling the situation in low signals which can be done with a help of signal quality index. The focus of this paper aims to solve those issues and develop a good algorithm that assesses the accurate level of anesthesia for most patients and it also aims to perform well when the signal quality is less.

2 Methods

The proposed methods are applied as the flow chart in Fig. 1. Initially, the denoised EEG data is passed through a bandpass filter where frequencies that fall within the specified range are taken and the remaining signal is rejected and reconstructed using the signal that falls within the range.

The band filtered signals are then passed through ICA algorithm where it tries to separate both the sources since both the channels may have data that may have been corrupted with noise and consists of artifacts since this is very normal with EEG data.

The independent channels are then taken to the next step where wavelet transformation is applied to the signals at level 5. This process is called feature extraction where a single channel is split up into 10 channels where adding up the extracted channels will produce the original channel. Wavelet transformation produces one approximation (a1) and one detail (d1) at each level and in this process 10 features are produced ranging from a1 to a5 and d1 to d5. Wavelet is done to control the oscillations that may occur in real-time data. Once the features are extracted, permutation entropy is performed on these features with a window size of 55 s since the input should match the dimensions of the BIS and usually, BIS recordings are done at every interval of 55 s.

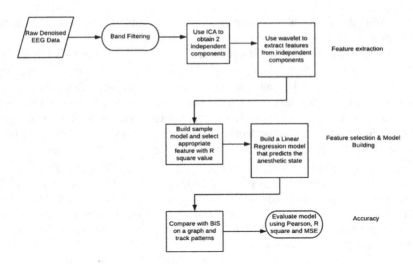

Fig. 1. Flow chart

The PE is calculated using the following algorithm [11]. Define the EEG signal $[x(i), i = 1, 2, ...]$ into a m-dimension space $X[x(i), x(i+L), ..., x(i+(m-1)L)]$ firstly, m is the number of dimension, L is the time delay. Secondly, sort the EEG series in the m dimension space in increasing sequence:

$$[x(i+(j_1-1)L) \leq x(i+(j_2-1)L) \leq ... \leq x(i+(j_m-1)L)] \tag{1}$$

$j_1, j_2, ...,$ and j_m show the new order of the series. For a m-dimension space, there are total $m!$ orders. Each $X[x(i), x(i+L), ..., x(i+(m-1)L)]$ reflects one of these '$m!$' orders. Assume the probabilities of each order are $P_1, P_2, ..., P_K$ respectively. According to the Shannon Entropy, the permutation entropy $PE(m)$ is calculated as follows:

$$PE(m) = -\sum_{j=1}^{K} P_j \ln P_j \tag{2}$$

To build the machine learning model, we can use all the 10 features generated that may lead to a robust model but the complexity of the model increases resulting in technical difficulties such as time delay. So, a process called as feature selection is applied where a sample machine learning model is built giving only one feature as an input and the R square is calculated for all the 10 features and the 2 most appropriate features that exhibited high R square value were selected to build the final model. The selected features were then given as input to the supervised Linear Regression Machine learning model where the label data was the BIS. The DoA index that was built in this process was plotted against the BIS index to check how well the proposed index was built compared to the BIS. The following equation is used to calculate the proposed index after the Linear regression model is built.

$$\text{New Index} = k(1) + k(2) * x1 + k(3) * x2 \tag{3}$$

Where k is derived from the factor b that is generated from the regress inbuilt function from the Matlab. The regress function is used for the linear regression model where factor b is generated which is a component used to calculate the proposed index. The factors x1 and x2 represent the input features that are fed into the model.

The model then calculates the proposed anesthetic state of a patient based on the inputs and the final model is then plotted against the BIS.

Finally, the proposed model is evaluated using Pearson Co-efficient, R square and root mean square error. The proposed model is also plotted against the BIS with the signal quality indicator so that it gives an overall visualization of the performance of both the proposed model and the BIS at low signal quality conditions.

3   Results

3.1   Data Pre-processing by ICA

The data sets that were used for this paper is time-series EEG signal. The sampling rate is 128 so the time series EEG data includes 128 readings for each minute. The data were obtained from 12 Adult Patients at the Toowoomba St Vincent's Hospital (7 used for training, 4 used for testing, 1 used for signal quality).

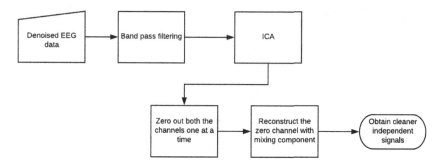

Fig. 2. Flow chart of data pre-processing

The two-channel raw EEG data were obtained through Quatro electrodes of BIS monitors which were placed diagonally on the forehead with electrode No. 1 at the centre of the forehead, electrode No. 4 directly above the eyebrow, No. 2 between No. 1 and No. 4, No. 3 on the temple, between the corner of the eye and hairline [12]. The EEG data pre-processing are processed in the flow chart as Fig. 2. The original denoised EEG data is passed onto the bandpass filter to reduce noise [13]. Bandpass filtering (BPF) takes the signal input and process the signals that fall within a frequency range (21.5 Hz–30 Hz) and rejects the remaining signal. The overall length of the signal is maintained by reconstructing the rejected parts with the help of the data that is present which falls under the frequency range. The resulting signal is a much more versatile signal with less noise. This process is done for both channels.

Independent component analysis is then performed on these band filtered signals to reduce noise and dependencies that may have occurred while extracting the signal from the brain. The algorithm takes both the signals as input and tries to separate each signal that is independent of each other and then channel 1 is replaced with zero and it is reconstructed using the mixing component that was originally used to separate the signals. The same process is repeated for channel 2 and the reconstructed signals are taken as input for the further process.

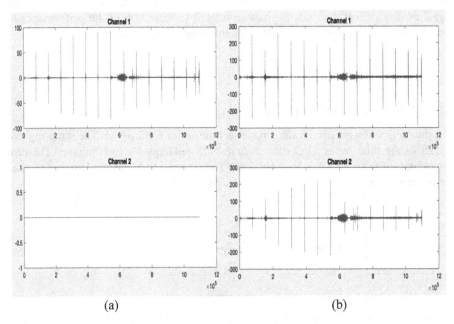

(a)   (b)

Fig. 3. Data pre-processing results (a) output after the ICA process (b) output after reconstructing

Figure 3(a) is the output after the ICA process where channel 1 is the output from ICA algorithm and channel 2 is also the output from the ICA algorithm but the values were replaced with zero. On looking at the signal of channel 1 closely compared to the original signal, the signal is smoother overall and the frequency range has reduced and the overall noise of the signal is reduced.

Figure 3(b) is the output after reconstructing the signal using the mixing component where channel 2 was obtained by mixing channel 1 and the mixing component that was originally used to separate the signals. The resulting output is a robust signal that has the same length and frequency range as the original signal and is also free from noise and corruption and artifacts sharing with the other signal. The same process is repeated for channel 1 where it is replaced with zero and reconstructed with the other channel. The output of these independent channels is then used for the next phase.

3.2   Feature Extraction and Feature Selection

The feature extraction and feature selection are processed in the flow chart as Fig. 4. The independent signals are then taken as input for this feature extraction process where a technique called wavelet is applied to these signals. The level of the wavelet that was used for this process was at 5 so the resulting set would have 10 features for each channel where 5 features are for the approximations and 5 features are for the details. It was found that db16 and sym8 wavelet were the most appropriate wavelet for EEG data. Once the features are extracted, permutation entropy is performed on these features with a window size of 55 since the input should match the dimensions of the BIS and usually BIS recordings are done at every interval of 55 s but there are 128 readings for a single minute for the EEG data.

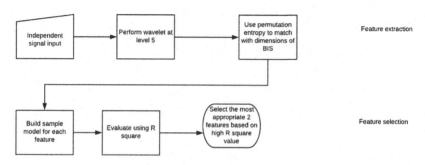

Fig. 4. Flow chart of feature extraction and feature selection

A sample model is then built on these extracted features where all the features were given as an input individually and 11 models were built and compared against the BIS and evaluated using R square value. The two Features that exhibited high R square value was selected as the final features that would be fed as an input to the machine learning model.

Take Patient 4 for example, the R square values are calculated by 11 signals which include the original signal, its a1 to a5 and its d1 to d5. The highest R square value is represented by the features a1 and d2. So, at the end of this step, it was decided that db16 wavelet transformation on Channel 1 with the features of a1 and d2 is the most appropriate features.

3.3   Model Building and Evaluation

The model building and evaluation are processed in the flow chart as Fig. 5. Once the appropriate features a1 and d2 that exhibited high correlation with the BIS were selected, a linear regression model as Eq. (4) was built on these features and the label input for the supervised machine learning model is the BIS readings.

$$\text{New Index} = 2074.81 - 2248.69 * \text{PEa1} + 1097.91 * \text{PEd2} \tag{4}$$

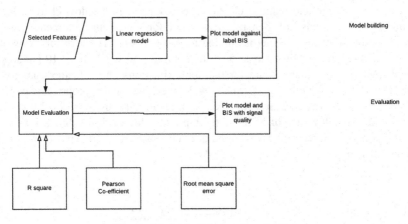

Fig. 5. Model building and evaluation

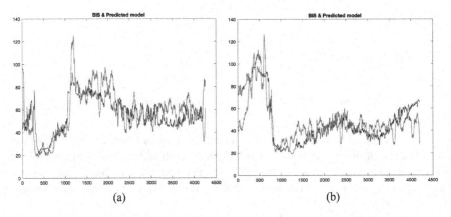

(a)   (b)

Fig. 6. Sample results of testing data, BIS (black) Predicted model (red) (a) patient 3 (b) patient 7 (Color figure online)

For combined training data of seven patients, the Pearson Co-efficient between the new index and BIS is 0.7308. Figure 6 shows the proposed model plotted against the BIS for the testing set of patients. The same feature extraction methods (a1 and d2) that were selected from the training set (patients 1, 2, 4 5, 6, 8, 9) were applied on the testing set (patients 3, 7, 10 and 11).

Take testing set patient 3 and patient 7 for example. The black plot shows the label values BIS and the red plot shows the new proposed index. From Fig. 6, it is clear that the proposed index gives a more stable value overall and holds well against the BIS. After plotting the BIS and Proposed index, the model is formally evaluated using R square, Root mean square error and Pearson Co-efficient. For the data set provided for this paper, the label BIS have a range of 0 to 130 and the RMS values for the observed values are shown below for all the training and testing sets that were used. For patient 3, the R square is 0.5324, Pearson Co-efficient is 0.7261 and MSE is 191.9214. For patient 7, the R square is 0.6311, Pearson Co-efficient is 0.7938 and MSE is 166.1439. For patient 10, the R square is 0.5602, Pearson Co-efficient is 0.7956 and MSE is 119.4157. For patient 11, the R square is 0.6603, Pearson Co-efficient is 0.7324 and MSE is 176.0294.

One more evaluation factor is taken into consideration where the proposed index is plotted against the BIS and the signal quality is taken into consideration now. This plot tells if the model performs well in low signal conditions where the BIS fails to provide any valid value.

The Fig. 7 shows the proposed model plotted against the BIS with the signal indicator for the patient 12. The blue line represents the signal indicator while the black line represents the BIS and the red line represents the proposed index. On looking closely at the second figure where the first figure was zoomed in to the situation where the quality of the signal drops it is found out that when the signal quality drops below 40 initially, the BIS tends to exhibit invalid or fuzzy values. A couple of factors may come into play for this behavior since BIS tends to start calculating the anesthetic level of patients after a certain amount of time but because the signal was a little low initially BIS flattens out and fail to provide any value. Meanwhile, the proposed model holds well in most signal conditions and gives an even and a smooth curve which tells us that the assessment of the patient's anesthetic state is quite stable.

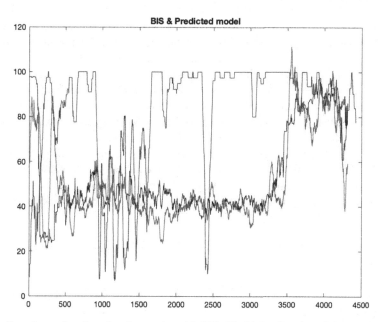

Fig. 7. Sample results of testing data patient 12, BIS (black) Predicted model (red) SQI (blue) (Color figure online)

4 Discussion

Because the feature PEa1 and PEd2 show a good linear relation with BIS, the linear regression is selected for creating the final index. Besides, comparing with other machine learning algorithms such as neural network, support vector machine and K-nearest neighbor algorithm, the computation complexity of a linear model is the lowest. Therefore, the predicted model based on the line model can reduce the time delay to some extent which is also a big challenge for real-time DoA assessment. Further work can be done by using different machine learning algorithms to improve the performance of the final index and using more patients' data for training to enhance the robustness of the predicted model.

5 Conclusion

In this paper, initially the denoised patient's data was taken and BPF was applied to the data to obtain a cleaner data that falls under the required range of frequency and frequencies that didn't fall in the range were removed and the overall length of the signal was still maintained because the remaining part of the signal was reconstructed using the signal that fell under the frequency range. ICA was then applied to the filtered signal and the aim was to remove the artifacts and dependencies both the signals may possess and both the channels were replaced with zero one at a time and reconstructed using the mixing component. The independent channels were then transformed using wavelet transformation where it was found that sym8 and db16 were the most appropriate types of wavelets for EEG data and db16 exhibited the highest R square value. By using the feature selection method, it was found that the channel 1 db16 output with the features of a1 and d2 was the most appropriate features that would help in assessing the accurate level of anesthesia. The features were fed as an input to the linear regression machine learning model with the label variable BIS and the output which was the proposed index was plotted against the BIS. It was found out that the proposed index holds well overall against the BIS and in certain cases performs better than the BIS which was validated in the high Pearson Co-efficient score and root mean square value which was consistent among all the patients that were used for training and testing. Finally, the proposed model was plotted against the BIS in low signal quality conditions and it was found out that when the signal quality is low initially, BIS provides invalid output and exhibits fuzzy and unreliable values due to its method of assessing anesthesia and thus it can be concluded that the proposed model is better than BIS.

References

1. Nguyen-Ky, T., Wen, P., Li, Y., Gray, R.: Measuring and reflecting depth of anesthesia using wavelet and power spectral density. IEEE Trans. Inf. Technol. Biomed. **15**(4), 630–639 (2011)
2. Bowdle, T.A.: Depth of anesthesia monitoring. Anesthesiol. Clin. **24**(4), 793 (2006)

3. Chen, D., Li, D., Xiong, M., Bao, H., Li, X.: GPGPU-aided ensemble empirical-mode decomposition for EEG analysis during anesthesia. IEEE Trans. Inf. Technol. Biomed. **14**(6), 1417–1427 (2010)
4. Myles, P.S., Leslie, K., McNeil, J., Forbes, A., Chan, M.T.V., B-Aware Trial Group: Bispectral index monitoring to prevent awareness during anaesthesia: the B-Aware randomised controlled trial. Lancet **363**(9423), 1757–1763 (2004)
5. Kumar, N.N., Reddy, A.G.: Removal of ECG artifact from EEG data using independent component analysis and S-transform. Int. J. Sci. Eng. Technol. Res. **5**, 712–716 (2016)
6. Bibian, S., et al.: Method and apparatus for the estimation of anesthetic depth using wavelet analysis of the electroencephalogram. U.S. Patent 7 373,198 (2008)
7. Wei, Q., et al.: Analysis of EEG via multivariate empirical mode decomposition for depth of anesthesia based on sample entropy. Entropy **15**(9), 3458–3470 (2013)
8. Nguyen-Ky, T., Wen, P., Li, Y.: An improved detrended moving-average method for monitoring the depth of anesthesia. IEEE Trans. Biomed. Eng. **57**(10), 2369–2378 (2010)
9. Kortelainen, J., Väyrynen, E., Seppänen, T.: Depth of anesthesia during multidrug infusion: separating the effects of propofol and remifentanil using the spectral features of EEG. IEEE Trans. Biomed. Eng. **58**(5), 1216–1223 (2011)
10. Nguyen-Ky, T., Wen, P.P., Li, Y.: Consciousness and depth of anesthesia assessment based on Bayesian analysis of EEG signals. IEEE Trans. Biomed. Eng. **60**(6), 1488–1498 (2013)
11. Jordan, D., Stockmanns, G., Kochs, E.F., Pilge, S., Schneider, G.: Electroencephalographic order pattern analysis for the separation of consciousness and unconsciousness: an analysis of approximate entropy, permutation entropy, recurrence rate, and phase coupling of order recurrence plots. Anesthesiol.: J. Am. Soc. Anesthesiol. **109**(6), 1014–1022 (2008)
12. Rezek, I., Roberts, S.J., Conradt, R.: Increasing the depth of anesthesia assessment. IEEE Eng. Med. Biol. Mag. **2**(26), 64–73 (2007)
13. Li, T., Wen, P., Jayamaha, S.: Anaesthetic EEG signal denoise using improved nonlocal mean methods. Australas. Phys. Eng. Sci. Med. **37**(2), 431–437 (2014)

A Preliminary Study of the Impact of Lateral Head Orientations on the Current Distributions During tDCS

Bo Song[1]([⊠]), Marilia Menezes de Oliveira[2], Shuaifang Wang[1],
Yan Li[1], Peng Wen[1], and Tony Ahfock[1]

[1] Faculty of Health, Engineering and Sciences, University of Southern
Queensland, Toowoomba, Australia
Bo.Song@usq.edu.au
[2] Faculty of Science, University of Sydney, Sydney, Australia

Abstract. This numerical study pre-validated the impact of lateral head orientations on the current distributions in the brain region during transcranial direct current stimulation (tDCS). A four-layer (scalp, skull, CSF, brain) real shape human head model was constructed with two electrodes configurations (C3-C4, C3-Fp2) and incremental downward displacement (0.5 mm) of the brain due to gravitational force. Sensitivity analysis was conducted on the influence of brain displacement during tDCS. Results of this preliminary study demonstrated that the cerebral current distribution was sensitive to the gravity-induced downward movement of the brain during tDCS, which suggested that lateral head orientations could be a new parameter to consider during tDCS and further research resources could be allocated in the realistic human head based studies to follow up this study. This finding should help both tDCS research and clinical trials to predict the stimulation result more precisely.

Keywords: tDCS · Human head modelling · Lateral head orientation

1 Introduction

In the past decades, tDCS has been broadly used as a research tool and therapeutic method for clinical mental disorders [1]. As a noninvasive brain stimulation technique, tDCS applies relatively weak current (0.5–2 mA) through the scalp to modulate the underlying cerebral functions and each session takes about 10–20 min [2, 3]. One of the challenges during tDCS applications is to understand how and where the current flows within the human brain. Simulations using human head models show promise for indicating the current distributions within the brain [4].

In the early stage of human head modelling, concentric spherical models were used broadly. A typical three-layer model has scalp, skull and brain. With the emerging of Magnetic Resonance Imaging and Diffusion Tensor Imaging, increasingly more researches focus on the realistic human head models. Many studies found that the cerebrospinal fluid (CSF) distribution had significant impacts on tDCS and many other neurophysiological modulation techniques [1, 5]. However, there is little research on

P. Liang et al. (Eds.): BI 2019, LNAI 11976, pp. 254–264, 2019.
https://doi.org/10.1007/978-3-030-37078-7_25

the impact of head orientations on the current distributions during tDCS. Since CSF acts as a buffer or cushion for the brain (as shown in Fig. 1), the brain may compress the CSF closer to the ground direction due to the downward shift caused by gravity. Such displacement was reported varying from 0.5 mm to 7.5 mm, averaging at 3.0 mm [6–8]. That is, the distribution of CSF is subject to change in the lateral head orientations as the CSF close to the ground side tends to be thinner than that of the other sides. This fact should be considered during precise applications of tDCS.

Therefore, this study was proposed to pre-validate the influence of lateral head orientations during tDCS before extensive resources are invested in the realistic human head based studies to quantitatively confirm and modelling such effect. This study numerically investigated the influences of brain displacements in two head orientations (left lateral and right lateral) with two different electrode montages (C3-C4, C3-Fp2) using 26 real shape human head models (as demonstrated in Table 1).

Table 1. Models applied with different montages, head orientations and brain displacements.

Group	Head orientation	Montage	Models notation	
			M[0]	M[n]
G1	Right	C3-C4	$M^{Control}_{C3-C4}[0]$	$M^{Right}_{C3-C4}[n]$
G2	Left			$M^{Left}_{C3-C4}[n]$
G3	Right	C3-Fp2	$M^{Control}_{C3-Fp2}[0]$	$M^{Right}_{C3-Fp2}[n]$
G4	Left			$M^{Left}_{C3-Fp2}[n]$

Notation $M^{Right}_{C3-C4}[n]$ represents a model applied with C3-C4 montage, right lateral head orientation and n mm brain shift towards the ground while $n \in \{0.5, 1, 1.5, 2, 2.5, 3\}$.

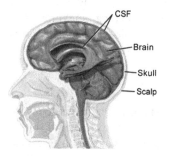

CSF
Brain
Skull
Scalp

Fig. 1. Four layers included in this study (Image modified from [9])

Table 2. Parameters assigned to all layers

Structure	Semi axis (mm)			Conductivity (S/m)
	a	b	c	
Skull	70	90	90	0.015
CSF	66	86	86	1.79
Brain	62	82	82	0.2
Scalp				0.43
Electrode	40 × 40			0.14

2 Methodology

The models used were developed based on the Rush and Driscoll human head model (R.D Model) [10]. However, several major changes were applied to make it a realistic shaped human head model. Specifically, CSF layer was introduced between the skull and brain layers. All the layers were modelled as concentric ellipsoids, rather than concentric spheres in R.D Model. The regions of interest (ROI) chosen was the whole surface of the brain layer and the distribution of current density in this surface was analyzed with different electrode montages, head orientations and brain displacements.

2.1 Quasi-static Approximation and Boundary Conditions

Since tDCS utilizes direct current, the Quasi-static approximation could be employed in the modelling process. Therefore, all head layers in this study were modelled as passive volume conductors and their intracellular volume-averaged current source was ignored, which complied with quasi-static Laplace equation [11]:

$$\nabla \cdot (-\sigma \nabla \phi) = 0 \tag{1}$$

where ϕ is the electric potential and σ is the electrical conductivity of a conductive medium. For isotropic volume conductors assumed in this study, ϕ is merely a scalar.

With the quasi-static approximation, the relationships between electric filed (E) and current density (J) of the medium can be defined in Eqs. (2) and (3) [12]:

$$E = -\nabla \phi \tag{2}$$

$$J = \sigma E. \tag{3}$$

For this study, inward current density was applied to anode and cathode under. As air-scalp boundary is generally considered as electric insulating, no current density or ground boundary condition ($\phi = 0$) was applied to the air-scalp boundary while the continual current was maintained in all interior tissue-tissue boundaries.

2.2 Model Construction and Stimulation Parameters

The electrical conductivity of Skull and CSF were obtained from literature reviews [13, 14]. The scalp conductivity was derived from the conductivity values of skin, fat and muscles [15]; Grey matter and white matter contributed to the brain layer conductivity [16, 17]. The size of all sponge electrodes was modelled as 4.0 cm × 4.0 cm. The sponge electrodes are typically soaked in saline solution in clinical trials. Therefore, all electrodes were assigned with the conductivity of the saline solution [17].

The layer of scalp was modelled from SAM Phantom while the dimensions of skull, CSF and brain were scaled with the same ratio as R.D Model to fit in the real shape model [7, 10]. The dimension and conductivity of all layers are listed in Table 2.

Montages C3-C4 and C3-Fp2 were chosen to simulate the primary motor cortex [2, 18]. The exposed scalp surface was assigned with ground potential and flux continuity was set to all inner tissue-tissue boundaries. The anode (C3) was applied with 1.25 A/m^2 normal inward current density while the cathode (C4, Fp2) was applied with -1.25 A/m^2 normal inward current density (2 mA current). Both montages and lateral head orientations were implemented in all four groups. Specifically, M[0] was the control model with no brain displacement while in M[0.5]–M[3.0] models, the displacements were 0.5 mm, 1.0 mm, 1.5 mm, 2 mm, 2.5 mm and 3 mm.

2.3 Finite Element Model Mesh and Data Computation

Using COMSOL Multiphysics software package, all the models with designed geometry properties (montage, brain displacement etc.) were constructed and assigned with proposed boundary conditions and other stimulation parameters. Then, physics controlled mesh was conducted to generate tetrahedral finite elements, which resulted in about 1.3 million elements in each model.

The surface of the brain was chosen as the only ROI and the distribution of current density on that surface was computed. For each model, three characteristic values were computed from its current density dataset. Namely, they were 99th percentile value, median value and threshold area (TA). To be specific, 99th percentile was used to avoid the computational instabilities. TA was defined as the percentage of surface area where the amplitude of current density was greater than 70% of its 99th percentile [12] (Fig. 2).

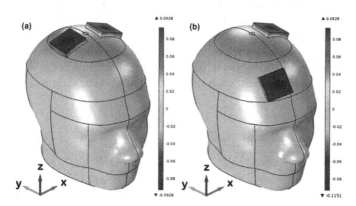

Fig. 2. Current distributions in the control models of C3-C4 (a) and C3-Fp2 (b) montages

3 Result

3.1 Analysis of Models with C3-C4 Montage

Three key feature values of G1 were described Fig. 3a (M[0] is the control model while M[0.5] represents $M_{C3-C4}^{Right}[0.5]$). As can be clearly observed, with the gradual increase of brain displacement, the values of 99th percentile and TA respectively increased about 6.5% and 21% while the median value almost remained the same (1.8×10^{-3}–

1.9×10^{-3} A/m^2). Since only a small proportion of current penetrates through the skull and other layers, the current density on the brain surface was scaled by Log_{10} and the normalized values were classified into eight bins (Bin1–Bin8). Figure 3b revealed that with the increase of brain shift, the fraction volume of normalized current density in peak range (Bin8) raised 24% while in the off-peak ranges of Bin7 and Bin4, it declined about 12% and 5%, respectively. As for the normalized current density in other bins, they slightly fluctuated but still remained at the same level as that of the control model.

As shown in Fig. 4a, the same trend could also be observed under the configuration of C3-C4 montage and the left lateral head orientation. To be specific, with the gradual increase of brain displacement, the 99th percentile and TA increased about 5.9% and 23.8% while the median value still stayed at the same level (1.8×10^{-3}– 1.9×10^{-3} A/m^2). A similar normalized current density distribution could be seen from Fig. 4b. The fraction volume of normalized current density in peak range (Bin8) increased about 24% and the scaled value of Bin7 and Bin4 declined about 13% and 4.9%. And the normalized current density in other bins almost remained at the same level. The tiny difference between the results of right lateral head orientation and left head orientation models was because that the geometry of this model, especially for the scalp layer, was not exactly symmetric.

These findings implied that under the configuration of C3-C4 montage, no matter which lateral head orientation was applied, the increase of brain shift could bring more current penetrated through the scalp and reached the target region of the brain surface. As a result, it gradually enhanced the stimulation effect on the primary motor cortex. However, as can be seen from G1 and G2 of Fig. 5, with the increase of brain displacement, the center of target region was shifting. G1 was configured with right lateral head orientation and the target region center was shifting from the brain cortex to the middle of C4 and brain cortex. On the contrary, the stimulation center was moving to the middle of C3 and brain cortex when left lateral head orientation was applied (G2).

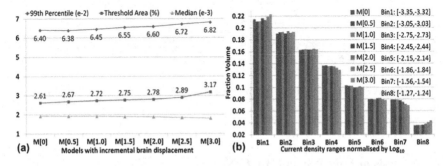

Fig. 3. Current density distributions on the brain surface of models with incremental brain displacements under C3-C4 electrode montage and right lateral head orientation (a: The key features of current density; b: The fraction volume distributions of Log_{10} normalized current density)

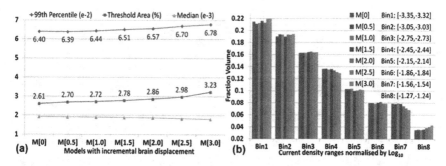

Fig. 4. Current density distributions on the brain surface of models with incremental brain displacements under C3-C4 electrode montage and left lateral head orientation (a: The key features of current density; b: The fraction volume distributions of Log_{10} normalized current density)

3.2   Analysis of Models with C3-Fp2 Montage

Figure 6a demonstrates the variation of 99th percentile, median and TA of G3. It is apparent to conclude that, with the moderate growth of brain displacement, the values of 99th percentile increased about 10% while TA declined approximately 12% while the median value declined a small portion (3%). Figure 6b reveals that with the increase of brain shift, the fraction volume of normalized current density in peak range (Bin8) declined about 27% while it increased about 15% and 6% in the off-peak Bin7 and Bin1. As for the normalized current density in other bins, they slightly fluctuated but could be still regarded at the same level because the changing gap was less than 1.9%, which could be considered as stable on the same level of the control model.

However, when the left head orientation was applied, with the increase of brain shift, a different trend could be observed from Fig. 7a. To be specific, when the brain displacement was modelled as a small quantity (0.5 mm, 1 mm), both the 99th percentile and TA remained the same level as the control model. Then, with the growth of brain displacement, the 99th percentile and TA were linearly increased approximately 9% and 7%, respectively. In terms of the median value, it declined an extremely small percentage (0.7%) and it could be considered as negligible when compared with the value of G3, which suggested that, with the increase of brain shift, the penetrated current reaching the brain surface remained almost the same level. From Fig. 7b, it is easy to discover that $M_{C3-Fp2}^{Left}[0.5]$ had a slightly higher normalized current density than the control model in the peak range (Bin8) and then with the increase of brain shift, the normalized current density in Bin8 declined about 16%. The normalized current density in off-peak ranges Bin7 and Bin1 increased 4% and 6% respectively while less than 2% variation gap could be observed for all the rest bins.

These findings implied that under the configuration of C3-Fp2 montage, with the increase of brain shift, huge differences could be observed in the current distributions when different lateral head orientations were applied. As demonstrated in Group3 (G3) and Group4 (G4) of Fig. 5, when the brain shift was increased under different lateral head orientations, the center of target region was also moving to different directions.

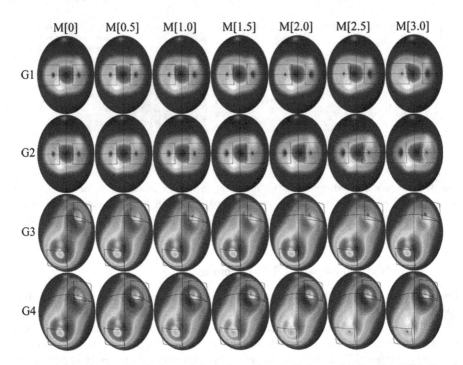

Fig. 5. Current density distributions of the brain surface in all the models listed in Table 1 and the right square pads in all four groups are the anode electrode placed at C3 location while the left ones are cathode electrode configured at C4 (G1, G2) and Fp2 (G3, G4) locations respectively

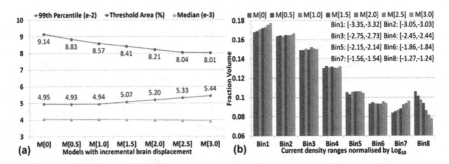

Fig. 6. Current density distributions on the brain surface of models with incremental brain displacements under C3-Fp2 electrode montage and right lateral head orientation (a: The distributions of three key features of current density; b: The fraction volume distributions of Log_{10} normalized current density)

Specifically, when no lateral head orientation was applied ($M_{C3-Fp2}^{Control}[0]$), two focuses could be identified in the target regions and they were the left primary motor cortex near the anode at C3 and right supra-orbital near the cathode at Fp2. However, when the right lateral head orientation was applied (G3), with the growth of brain shift, the

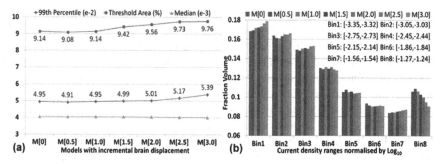

Fig. 7. Current density distributions on the brain surface of models with incremental brain displacements under C3-Fp2 electrode montage and left lateral head orientation (a: The distributions of three key features of current density; b: The fraction volume distributions of Log_{10} normalized current density)

focus at left primary motor cortex was compromised and at the same time, the other focus at right supra-orbital was strengthened and this local stimulation enhancement was achieved at a cost of the deterioration of the global TA. On the contrary, when the left lateral head orientation was applied (G4), the focus at the right supra-orbital was weakened and the focus at the left primary motor cortex was boosted. As a result, the local stimulation of left primary motor cortex was promoted though this promotion was not as strong as it in G3. With the increase of brain shift, the global TA in G4 was not undermined, except when the brain shift was set to 0.5 mm. But it was soon recovered and finally surpassed the original level in the control model ($M_{C3-Fp2}^{Control}[0]$), which was significantly higher than the TA in G3.

4   Discussions and Conclusions

This study has investigated the distribution of current density on the brain surface under the different configurations of electrode montages (C3-C4 and C3-Fp2), lateral head orientations (left and right) and brain displacements (0 mm–3.0 mm) by constructing a four-layer real shaped human head model. The experiment results and interpretations were focused on revealing the influence of brain shift on the distribution of current density on the brain surface, which were carried out by analyzing three feature values (99th percentile, median and TA) and the distribution of $\log_{10}$ normalized current density in four groups (G1–G4).

4.1   Comparison of Two Electrode Montages

Two electrode montages (C3-C4 and C3-Fp2) were investigated in this study, which are frequently used for the enhancement of motor performance of the non-dominant hand (C3-C4) and primary motor cortex of emotional and psychomotor functions (C3-Fp2) [19–22]. The simulation result derived from the control models of both montages suggested that the peak current density discovered on the brain surface was depended on the placement of electrodes. For symmetric electrode placement of C3-C4, the maximum current density was found right between the electrodes (M[0] of G1 and G2) while, in terms of asymmetric electrode montage of C3-Fp2, two peaks of current density could be identified beneath the adjacent corners of both electrodes (M[0] of G3 and G4). This finding corresponded to previous studies [4, 23, 24]. Besides, by comparing the median and TA values, more current could penetrate the out tissue layers (scalp, skull and CSF) and reach the brain layer under the C3-Fp2 montage due to that the inter-electrode distance in such montage was much bigger than it in C3-C4 montage, which decreased the shunting phenomenon of the current through the out layers [25]. However, the close inter-electrode distance in C3-C4 montage brought more focal current in the target region as its 99th percentile value was 1.3 times of it in C3-Fp2.

4.2   Variation of Brain Displacement

To simulate the downwards displacement of the brain caused by gravity under different electrode montages and lateral head orientations, incremental increase of the brain shift was applied as previous intraoperative studies reported a varied range of brain displacement [6, 26]. From the result of G1 and G2 with the electrode configuration of C3-C4, longer brain displacement could bring stronger stimulation effect on the primary motor cortex and the focus of target region also slightly shifted downwards to the ground direction. However, in terms of C3-Fp2 electrode configuration, the asymmetric electrode location determined different influences brought by the brain displacement under different lateral head orientations. In G3, higher brain shift caused a shrink of stimulation region while the area was fluctuating in G4 and determined by the extent of brain shift, which makes its application tricky for the future clinical trials as customized treatment is needed to cope this situation. Still, in both lateral head orientation under C3-Fp2, higher brain shift brought local enhancement of current density and this could be an important factor to consider during tDCS treatments.

4.3   Conclusions

This study constructed four groups of models with different electrode montages and lateral head orientations to investigate the influence lateral head orientations on the current density distribution during tDCS. The results demonstrated two major outcomes. Firstly, the downward movement of the brain made itself closer to one electrode (anode or cathode) and enhanced the stimulation of the brain region beneath that electrode at a cost of undermining the stimulation of the other electrode. Secondly, the overall current penetrated and current distributions on the majority of other brain regions remained almost unchanged at the same time. Therefore, the impact of lateral head orientations discovered in this study could be another factor to consider during the precise applications of tDCS for both researchers and clinical staffs. Meanwhile, further realistic human head modelling studies could be proposed to quantitatively confirm and model the impact of lateral head orientations during tDCS.

References

1. Sadleir, R.J., Vannorsdall, T.D., Schretlen, D.J., Gordon, B.: Transcranial direct current stimulation (tDCS) in a realistic head model. Neuroimage **51**, 1310–1318 (2010)
2. DaSilva, A.F., Volz, M.S., Bikson, M., Fregni, F.: Electrode positioning and montage in transcranial direct current stimulation. J. Vis. Exp. (2011)
3. Wagner, T., Valero-Cabre, A., Pascual-Leone, A.: Noninvasive human brain stimulation. Ann. Rev. Biomed. Eng. **9**, 527–565 (2007)
4. Wagner, T., Fregni, F., Fecteau, S., Grodzinsky, A., Zahn, M., Pascual-Leone, A.: Transcranial direct current stimulation: a computer-based human model study. Neuroimage **35**, 1113–1124 (2007)
5. Miranda, P.C., Lomarev, M., Hallett, M.: Modeling the current distribution during transcranial direct current stimulation. Clin. Neurophysiol. **117**, 1623–1629 (2006)
6. Letteboer, M.M.J., Willems, P.W., Viergever, M.A., Niessen, W.J.: Brain shift estimation in image-guided neurosurgery using 3-D ultrasound. IEEE Trans. Biomed. Eng. **52**, 268–276 (2005)
7. Shahid, S., Wen, P., Ahfock, T., Leis, J.: Effects of head geometry, coil position and CSF displacement on field distribution under transcranial magnetic stimulation. J. Med. Imaging Health Inform. **1**, 271–277 (2011)
8. Bijsterbosch, J.D., et al.: The effect of head orientation on subarachnoid cerebrospinal fluid distribution and its implications for neurophysiological modulation and recording techniques. Physiol. Meas. **34**, N9 (2013)
9. https://en.wikiversity.org/wiki/File:Blausen_0216_CerebrospinalSystem.png
10. Rush, S., Driscoll, D.A.: Current distribution in the brain from surface electrodes. Anesth. Analg. **47**, 717–723 (1968)
11. Bai, S., Loo, C., Dokos, S.: A review of computational models of transcranial electrical stimulation. Crit. Rev™. Biomed. Eng. **41**, 21–35 (2013)
12. Parazzini, M., Fiocchi, S., Rossi, E., Paglialonga, A., Ravazzani, P.: Transcranial direct current stimulation: estimation of the electric field and of the current density in an anatomical human head model. IEEE Trans. Biomed. Eng. **58**, 1773–1780 (2011)
13. Baumann, S.B., Wozny, D.R., Kelly, S.K., Meno, F.M.: The electrical conductivity of human cerebrospinal fluid at body temperature. IEEE Trans. Biomed. Eng. **44**, 220–223 (1997)
14. Oostendorp, T.F., Delbeke, J., Stegeman, D.F.: The conductivity of the human skull: results of in vivo and in vitro measurements. IEEE Trans. Biomed. Eng. **47**, 1487–1492 (2000)
15. Holdefer, R., Sadleir, R., Russell, M.: Predicted current densities in the brain during transcranial electrical stimulation. Clin. Neurophysiol. **117**, 1388–1397 (2006)
16. He, B.: Modeling and Imaging of Bioelectrical Activity, pp. 281–316. Springer, Boston (2004). https://doi.org/10.1007/978-0-387-49963-5
17. Datta, A., Bansal, V., Diaz, J., Patel, J., Reato, D., Bikson, M.: Gyri-precise head model of transcranial direct current stimulation: improved spatial focality using a ring electrode versus conventional rectangular pad. Brain Stimul. **2**, 201–207 (2009). e201
18. Song, B., Wen, P., Ahfock, T., Li, Y.: Numeric investigation of brain tumor influence on the current distributions during transcranial direct current stimulation. IEEE Trans. Biomed. Eng. **63**, 176–187 (2016)
19. Boggio, P.S., et al.: Enhancement of non-dominant hand motor function by anodal transcranial direct current stimulation. Neurosci. Lett. **404**, 232–236 (2006)

20. Boggio, P.S., Nunes, A., Rigonatti, S.P., Nitsche, M.A., Pascual-Leone, A., Fregni, F.: Repeated sessions of noninvasive brain DC stimulation is associated with motor function improvement in stroke patients. Restor. Neurol. Neurosci. **25**, 123–129 (2007)

21. Utz, K.S., Dimova, V., Oppenländer, K., Kerkhoff, G.: Electrified minds: transcranial direct current stimulation (tDCS) and galvanic vestibular stimulation (GVS) as methods of non-invasive brain stimulation in neuropsychology—a review of current data and future implications. Neuropsychologia **48**, 2789–2810 (2010)

22. Koenigs, M., Ukueberuwa, D., Campion, P., Grafman, J., Wassermann, E.: Bilateral frontal transcranial direct current stimulation: failure to replicate classic findings in healthy subjects. Clin. Neurophysiol. **120**, 80–84 (2009)

23. Peterchev, A.V., Rosa, M.A., Deng, Z.-D., Prudic, J., Lisanby, S.H.: ECT stimulus parameters: rethinking dosage. J. ECT **26**, 159 (2010)

24. Datta, A., Elwassif, M., Battaglia, F., Bikson, M.: Transcranial current stimulation focality using disc and ring electrode configurations: FEM analysis. J. Neural Eng. **5**, 163 (2008)

25. Faria, P., Hallett, M., Miranda, P.C.: A finite element analysis of the effect of electrode area and inter-electrode distance on the spatial distribution of the current density in tDCS. J. Neural Eng. **8**, 066017 (2011)

26. Hu, J., et al.: Intraoperative brain shift prediction using a 3D inhomogeneous patient-specific finite element model. J. Neurosurg. **106**, 164–169 (2007)

Identification of Stress Impact on Personality Density Distributions

Brendan Lys[1], Xiaohui Tao[1(✉)], Tony Machin[2], Ji Zhang[1], and Ning Zhong[3]

[1] School of Sciences, University of Southern Queensland,
Toowoomba, Australia
{Brendan.Lys,Xiaohui.tao,Ji.Zhang}@usq.edu.au
[2] School of Psychology and Counselling,
University of Southern Queensland, Toowoomba, Australia
Tony.Machin@usq.edu.au
[3] Department of Life Science and Informatics,
Maebashi Institute of Technology, Maebashi, Japan
zhong@maebashi-it.ac.jp

Abstract. The high cost of stress both at the individual and societal levels is well documented. This study seeks to explore a new approach to the detection of individuals suffering from high levels of stress, through the analysis of changes in personality density distributions in relation to stress. The proposed approach is to gain personality profile information from text - building density distributions from these profiles, and using this same text to carry out stress analysis. The density distributions are then further analysed to explore the potential to identify density distribution shape changes in relation to stress.

1 Introduction

The impact of stress ranges from individual through to societal levels. The recognition of this impact has and continues to position stress as a subject for active academic enquiry. Within the context of organisations and employee stress, challenges include the timely identification of employees potentially suffering from stress, and the individual nature of stress in relation to individual appraisal of environmental demands. Findings from the study of personality coupled with Machine Learning Personality Assessment, may hold potential for both individualising stress measurement and allowing timely identification of potentially negative levels of stress at the individual employee level.

The Big Five Factors of Personality (referred to as the Big Five within this document) are considered by some to be one of the major achievements of the Psychological discipline [1]. Critical to this study is a subsequent finding [2], that of how the Big Five exhibit themselves over time (Fig. 1).

Fleeson [2] described the Big Five as displaying density distributions. As the below figure shows, while this individual can be accurately described as being introverted, they also exhibit levels of both introversion and extroversion - albeit

P. Liang et al. (Eds.): BI 2019, LNAI 11976, pp. 265–272, 2019.
https://doi.org/10.1007/978-3-030-37078-7_26

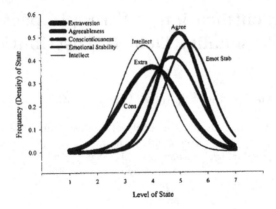

Fig. 1. Distribution of personality states over a two week period [2].

at lower frequencies. That is to say, within the study of personality both trait based and social cognition based approaches to personality have a place. Trait based descriptions of personality describe the mean personality, while depending on the situation (filtered through social-cognition processing) we may exhibit personality traits to the left or right of the mean. Potentially then, under stress, through social cognition processes the density distribution of trait expression may move away from a bell curve.

Central to this research then, is the following question:

Can individual experiences of excessive stress be seen in changes to their individual trait based density distributions?

The proposed research will contribute to the following three areas:

Knowledge Advancement. Currently personality trait movement have been described as density distributions [2], however the impact of stress on these density distributions is not understood, this research will advance our knowledge in this area.

Methodological Contribution. Psychologists using machine learning and knowledge discovery techniques in conjunction with statistical approaches are currently in the minority, with the majority continuing to focus solely on statistical analysis. This research aims to add to the increasing body of work, which is demonstrating the application of machine learning within the psychological discipline, with the aim of encouraging a broader consideration and potentially adoption of these approaches.

Community Benefit. The individual and financial cost of stress is well documented. Through the testing of a method to identify employees experiencing excessive stress, the potential is there to reduce negative impacts of excessive stress.

The paper is organized as following. In next section the works related to the topic are discussed, including those about integrated personality theory and Machine Learning personality assessment. After that, Sect. 3 reports our work on the topic and Sect. 4 presents some related scenarios with deep understanding of personality through Machine Learning techniques. Finally, Sect. 5 gives the final remarks to the work.

2 Related Work

2.1 Integrated Personality Theory

At the turn of the 21st century, the field of personality research was engaged in continuing debate between trait based (Big Five) and social cognition based theories of personality [3]. The extremes of each perspective on these approaches to theorising personality can be summarized as follows. The trait-based approach views personality as fixed and does not change. While at the other extreme, the social cognition approach views personality as changing in response to both internal and external stimulus. While both sides had acknowledged "room" for the other [1], lacking was the empirical evidence to provide a common ground for both approaches to be united. This evidence or common ground was found in 2001 [2], through the demonstration that personality trait expression forms bell curve like distributions over time (see Fig. 1).

Thus, personalities move dynamically around a mean in response to internal and external stimulus, therefore taking on the properties of social cognition theories of personality, while being describable in terms of traits. For example an individual whose personality can be described as introverted, may display an increased level of extroversion while giving a lecture. Similarly, an individual whose personality can be described as extroverted, may display a decreased level of extroversion while attending that same lecture as a participant. This finding gave rise to the earlier stated research question. The movement of personality trait displays in response to internal and external stimulus, may provide insight into stress being experienced by the individual.

2.2 Machine Learning Personality Assessment

The application of machine learning techniques to personality is an active and growing research area. Vinciarelli et al. [4] notes within their 2014 survey that the search term "personality" returned 6127 and 1033 hits on ACM Digital Library and IEEE XPlore respectively. The majority of early machine learning inquiry into personality was methodology based - that is how can we infer or measure personality without using traditional pen and paper tests [5]. The results of this methodological focus have certainly born fruit with the emergence and validation of Machine Learning Personality Assessment (MLPA) approaches [6], allowing researchers and lay people alike to gain quantitative individual personality insights from digital sources such as documents.

While a multitude of different digital sources have been utilised for gaining these personality insights: including analysis of social media profile pictures; GitHub contributions; and Facebook likes, text based analysis has emerged as one of the major sources for gaining an individuals personality profile [7]. And it is text data that this study will also use as its raw material to extract personality profiles, based on the application of MLPA. Specifically we will be testing our research question on a sample of emails from a large organisation.

3   Research Framework

This study seeks to explore density distributions of personality traits, in combination with stress measures of the same text. It will accomplish this through gaining personality profile information from text, and using this same text will carry out stress analysis. With stress scores being compared to personality profile shifts away from the mean through correlation analysis, to determine the relationship between density distribution movement and stress.

3.1   Data Understanding

Fundamental questions asked during the data understanding phase of a data mining project include: what data is required?; can we access the required data?, and is the data suitable for the evaluation of our hypotheses?. The response to the first of these questions is straightforward, essential for this research are personality profiles and stress scores of the same subjects collected over a period of time. The remaining questions present methodological challenges when asked of traditional questionnaire driven data gathering. For example gaining longitudinal data presents the challenge of gaining and retaining study subjects [8], additionally the relationship between the personality profile and the stress score may be less than direct.

3.2   Modelling

Having gained k personality profiles and stress score samples, we then turn to analysing these to test for correlations between density distribution changes within the personality profiles and accompanying stress scores. Furthermore, we will be training and testing models to predict density distribution movement in relation to stress scores. We will be adopting k cross validation, where the personality profile set of the individual and attached stress score are split across k groups, where k−1 samples are used to train the model, with the remaining personality profile used to test the model. This process is repeated k times, so at the end of the process each profile within the set has been used to train and test models.

Different Machine Learning techniques will be trailed for modelling. Specifically, Recurrent Neural Networks such as Long short-term memory networks (LSTMs) will be first tested. LSTMs were introduced by Hochreiter and Schmidhuber in 1997 [9] and were then refined and popularized in many following work. LSTMs are explicitly designed to avoid the long-term dependency problem and have the capacity of learning a model from samples collected over a long period of time. Such a design is also confirmed by many state-of-the-art works that adopted LSTMs for behavioral learning and human personality. For example, Majumder et al. [10] presented a deep learning based method using LSTMs for determining the author's personality type from text; Tandera et. al. [11] used LSTMs to develop a system to predict personality from Facebook users. We will firstly trail LSTMs with its variations to learn the model of density distribution and stress interaction.

4 Scenarios

Stress at work is recognised widely as a global issue [12], consider however that while a global issue, stress is an outcome at the individual level. [13] operationalises stress as resulting from the individuals appraisal that a particular environmental demand(s) are about to tax their individual resources. The stress experienced between two individuals in similar situations therefore can be wildly different. The stress experienced in providing CPR to a young child, will be different between a trained paramedic and an untrained parent for example. Also consider that not all stress is bad [14], indeed some levels of stress are positive in terms of performance benefits. As a trained paramedic, the stress experienced in applying life saving techniques may lift performance, while as an untrained parent the need to resuscitate their child may over come their individual resources and result in stress which negatively impacts their ability to perform.

Within the location of the workplace, we are faced with the balancing act of reducing stress which overcomes the resources of the individual, yet providing some stress to promote performance, while having potentially limited insight into the individuals resources. Stress is the outcome of a process which is cumulative, pressures from non-work sources may reduce the resources available to the individual, what was possible last month may this month exceed the available resources due to other demands. The criticality of the manager subordinate relationship in managing and coping with stress has been identified in previous studies [12]. However not all managers or indeed employers are equal in recognising stress or their impacts on mental health [12].

The above then presents an interesting challenge. At the individual employee level we have an unknown (and potentially constantly shifting) quantity in terms of individual resources, while at the employer/manager level we have varying ability to recognise harmful stress levels. What is needed is a system of measurement which is individualised, operating in the background to constantly assess stress, and to provide actionable insights back to the relevant parties. Machine Learning Personality Assessment in combination with personality theory, may be able to fill this gap.

Consider the hypothetical case of Frank, Frank is a knowledge worker in a large firm, and has been a solid performer within his role for the past five years. Unfortunately Franks marriage is breaking down, leading to high levels of stress within Franks home life. Franks colleagues have noticed individually that Frank seems a little less engaged lately, individually however this is not given much thought as everyone is busy. Frank hasn't shared his problems at home with anyone at work, so his potential decrease in individual resources isn't recognised by this employer. Frank begins to take sick leave over a period of time, an afternoon here, a few days there. Franks work begins to suffer, which comes to the attention of this manager, Frank is called into a meeting to discuss his employers concerns regarding his performance and attendance at work.

The above scenario could however be changed through measuring stress in almost real time. Digital means of communication, with a focus on email communication, are prolific within modern organisations. These digital breadcrumbs can be sifted through automatically by knowledge discovery techniques such as MLPA, providing individual specific measurement coupled with comparisons to that individuals mean personality profile and estimated density distribution overtime. An increased frequency of profile readings outside of the expected distribution will provide information that the individual may be operating outside of their ideal stress range. The following two distributions provide an example of such measurement. The first is the expected distribution informed by personality theory [2] and through historical analysis - in this example simplified to one density distribution rather than five (Figs. 2 and 3):

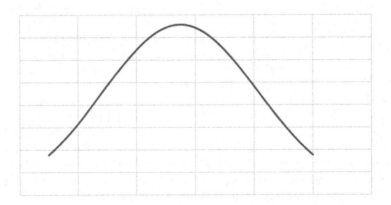

Fig. 2. Expected frequency distribution

While the second distribution (Fig. 3) is the actual distribution found on current analysis of Franks emails.

Such information can then be provided to registered health professionals, who can make informed judgements about the individual and decide on the appropriate course of action - which will vary between action being taken through to no action being required.

Taking into account the proliferation of digital communication such as emails, let us revisit Frank and his situation. Through the cumulative process of his marriage breaking down and his work demands, Frank's individual resources have been reduced. Frank experiences environmental demands that exceed his individual resources, work tasks that were once relatively simple are now overly

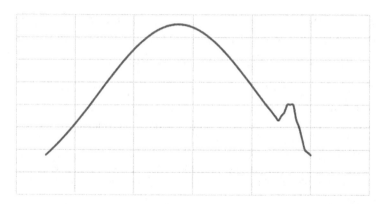

Fig. 3. Actual frequency distribution

challenging and stressful. Through comparing Franks personality density distribution change over time, we can see that Frank for reasons unknown is experiencing levels of stress which are potentially negatively impacting his performance. Franks personality profile information is forwarded automatically to a registered health practitioner to make an assessment and determine a further course of action. A call is made to Franks manager, with the resulting conversation revealing that Frank has been taking some sick leave, and maybe isn't himself. A meeting is called, however unlike the previous scenario this meeting isn't about Franks performance, this meeting is about Frank and checking in with him that he's okay.

Through appreciating that Franks experience of stress may have changed, as suggested by his change in personality profile distribution, the basis of the conversation above has changed from a potentially confrontational performance management focus to one which is focused on employee well being.

5   Conclusions

The identification of potential signs of stress, and passing these along to qualified health practitioners for further evaluation, has the potential to reduce the individual and societal costs of stress. Through using personality density distributions, we may be able to overcome the challenge of differing levels of individual resources in relation to the appraisal of environmental demands.

References

1. Fleeson, W., Jayawickreme, E.: Whole trait theory. J. Res. Pers. **56**, 82–92 (2015)
2. Fleeson, W.: Toward a structure-and process-integrated view of personality: traits as density distributions of states. J. Pers. Soc. Psychol. **80**(6), 1011 (2001)
3. Fleeson, W., Noftle, E.: The end of the person-situation debate: an emerging synthesis in the answer to the consistency question. Soc. Pers. Psychol. Compass **2**(4), 1667–1684 (2008)

4. Vinciarelli, A., Mohammadi, G.: A survey of personality computing. IEEE Trans. Affect. Comput. **5**(3), 273–291 (2014)
5. Wright, A.G.: Current directions in personality science and the potential for advances through computing. IEEE Trans. Affect. Comput. **5**(3), 292–296 (2014)
6. Farnadi, G., et al.: Computational personality recognition in social media. User Model. User-Adapt. Interact. **26**(2), 109–142 (2016). https://doi.org/10.1007/s11257-016-9171-0
7. Kaushal, V., Patwardhan, M.: Emerging trends in personality identification using online social networks—a literature survey. ACM Trans. Knowl. Discov. Data (TKDD) **12**(2), 15 (2018)
8. Cumming, J.J., Goldstein, H.: Handling attrition and non-response in longitudinal data with an application to a study of Australian youth. Longitud. Life Course Stud. **7**(1), 53–63 (2016)
9. Schmidhuber, J., Hochreiter, S.: Long short-term memory. Neural Comput. **9**, 1735–1780 (1998)
10. Majumder, A.G.N., Poria, S., Cambria, E.: Deep learning-based document modeling for personality detection from text. IEEE Intell. Syst. **32**, 74–79 (2017)
11. Tandera, T., Hendro, Prasetio, Y.L.: Personality prediction system from Facebook users. Proc. Comput. Sci. **116**, 604–611 (2017)
12. Dewe, P.J., O'Driscoll, M.P., Cooper, C.: Coping with Work Stress: A Review and Critique. Wiley, Hoboken (2010)
13. Holroyd, K.A., Lazarus, R.S.: Stress, coping and somatic adaptation. In: Handbook of Stress: Theoretical and Clinical Aspects, pp. 21–35 (1982)
14. McGonigal, K.: The Upside of Stress: Why Stress Is Good for You, and How to Get Good at It. Penguin, New York (2016)

Author Index

Printed in the United States
By Bookmasters